Achieving ISO/IEC 20000
Making metrics work

The 'Achieving ISO/IEC 20000' series

This publication is the third in a series of ten publications related to ISO/IEC 20000. Each publication provides advice on different aspects of ISO/IEC 20000. The books in the 'Achieving ISO/IEC 20000' series are:

Management decisions and documentation (BIP 0030)

Why people matter (BIP 0031)

Making metrics work (BIP 0032)

Managing end-to-end service (BIP 0033)

Finance for service managers (BIP 0034)

Enabling change (BIP 0035)

Keeping the service going (BIP 0036)

Capacity management (BIP 0037)

Integrated service management (BIP 0038)

The differences between BS 15000 and ISO/IEC 20000 (BIP 0039)

This series provides practical guidance and advice on introducing IT service management best practice in accordance with ISO/IEC 20000. More details on the content of each publication are given in *Books in the 'Achieving ISO/IEC 20000' series*, at the end of this book.

Although security issues are covered in ISO/IEC 20000, the 'Achieving ISO/IEC 20000' series does not cover security requirements. Information on security can be found in the BSI publications that are listed in the *Bibliography* in Appendix B.

Other publications

BSI also publishes:

A managers' guide to service management (BIP 0005) is intended for managers who are new to support services or who are faced with major changes to their existing support facility. This book takes the form of informative explanations, guidance and recommendations.

IT service management – Self-assessment workbook (BIP 0015) is an easy to use checklist that complements ISO/IEC 20000 and is designed to assist an organization's internal assessment of their services and the extent to which they conform to the specified requirements in ISO/IEC 20000.

Achieving ISO/IEC 20000
Making metrics work

Dr Jenny Dugmore and Shirley Lacy

Business Information

First published in the UK in 2005

Second edition published in the UK in 2006

by
BSI
389 Chiswick High Road
London W4 4AL

Typeset in Frutiger by Typobatics Limited
Printed in Great Britain by MPG Books Limited, Bodmin, Cornwall

British Library Cataloguing in Publication Data
A catalogue record for this book is available from the British Library

ISBN 0-580-47460-7

Contents

Foreword

ISO/IEC 20000 is about understanding how service management contributes to the customer's and service provider's business needs and strategic objectives. Without best practice service management, services can be unreliable, expensive and the cause of customer dissatisfaction.

This publication is the third in the 'Achieving ISO/IEC 20000' series. It describes the role of the service reporting process in ISO/IEC 20000. Without best practice service reporting, service management may not be effective, regardless of how close other processes are to the best practices, and regardless of the skill levels of the support staff and managers.

Service reporting, based on a sound set of metrics, enables a manager to effectively monitor services and processes, and to identify and implement improvements. This publication focuses on the planning and design of effective metrics and service reporting. It also identifies the features, uses and common pitfalls of metric production, and explains the role of metrics in the management of interfaces and their role in service management as a whole. Example metrics used to manage individual processes are given in the other publications in the 'Achieving ISO/IEC 20000' series.

Dr Jenny Dugmore

Acknowledgements

This book has been produced with the input and assistance of people involved in the practical aspects of delivering services across all sectors. We would like to thank them for sharing their views and providing constructive criticism, case studies and practical techniques.

The authors would like to thank the following for their support and assistance during the writing of this book: Barbara Eastman; Colin Hamilton of Renard Consulting Ltd; Dave Cuthbertson; Don Page of Marval; Hella Schrader of the Financial Times; Lynda Cooper of Fox IT; Maggie Kneller; R. Srinivasan of Wipro Technologies; Ro Gorell; Sonia Yusuf; and Sharon Hampton of Serco Solutions.

Finally, we would like to thank Simone Levy and Kieran Parkinson of BSI for their support, helpful suggestions, tact and patience during the production of the 'Achieving ISO/IEC 20000' series.

INTRODUCTION

What is ISO/IEC 20000?

ISO/IEC 20000 is the first IT service management process standard to be produced by the International Organization for Standardization (ISO), and it is based on the knowledge and experience gained by experts working in the field.

ISO/IEC 20000 was produced by Technical Committee ISO/IEC JTC 1/SC 7, *Software and system engineering*, and was based on BS 15000, which was produced by BSI Technical Committee BDD/3, *Information services management*.

ISO/IEC 20000 is in two parts.

- ISO/IEC 20000-1 is a specification containing requirements that must be met in order to achieve ISO/IEC 20000.
- ISO/IEC 20000-2 is a code of practice on how to achieve the requirements in ISO/IEC 20000-1.

The requirements in Part 1 (i.e. ISO/IEC 20000-1) are applicable to service providers of all sizes and types, regardless of whether the organization is public or private sector, internal or external. The recommendations in Part 2 (i.e. ISO/IEC 20000-2) are optional approaches to achieving the requirements in Part 1. Although optional, the recommendations are also practical and proven methods that are normally appropriate.

The purpose of ISO/IEC 20000

ISO/IEC 20000 provides the basis for assessing whether service providers have best practice, reliable, repeatable and measurable processes applied consistently across their organization. As a process-based standard the requirements are independent of organizational structure or of the tools used to automate the service management processes. ISO/IEC 20000-1 provides the basis for formal certification schemes and other audits.

The 'Achieving ISO/IEC 20000' series

The 'Achieving ISO/IEC 20000' series is designed to explain the requirements of ISO/IEC 20000. An abstract of the ISO/IEC 20000 clauses that are most relevant to the topic of 'Making metrics work' is given in Appendix A. Also, Table 1 provides a clause-by-clause guide to the content of each of the books in the 'Achieving ISO/IEC 20000' series.

This third publication in the 'Achieving ISO/IEC 20000' series focuses on the importance of a service provider understanding and producing the types of metric that are required for the management of a service. It emphasises the need for metrics designed to meet the target audiences needs, whether the audience is a customer, a supplier or one of the service providers own managers. It covers the principles of top-down design and includes practical hints, case studies and examples.

This publication also includes case studies, examples and other useful information on making metrics work for a service provider aiming to achieve ISO/IEC 20000. Many of the case studies and examples in this publication have been based on experiences with BS 15000, which was the precursor to ISO/IEC 20000.

Additional advice

Service providers aiming to achieve ISO/IEC 20000 may find it useful to seek advice on best practice, the qualifications available for individual service management professionals and ISO/IEC 20000 certification. Details of these can be found via the web pages cited in Appendix B.

Table 1 – Clause-by-clause guide to the 'Achieving ISO/IEC 20000' series

ISO/IEC 20000 clause	BIP 0030	BIP 0031	BIP 0032	BIP 0033	BIP 0034	BIP 0035	BIP 0036	BIP 0037	BIP 0038	BIP 0039
Terms and definitions	■									■
Management responsibility	■								■	■
Documentation requirements		■								■
Competence, awareness and training		■							■	■
Planning and implementing service management										■
Plan – Do – Check – Act cycle										■
Planning and implementing new or changed services										■
Service level management				■						■
Service reporting			■							■
Service continuity and availability management							■			■
Budgeting and accounting for IT services					■					■
Information security management	See BSI publications on information security management in the BIP 0070 series									
Capacity management								■		■
Business relationship management				■						■
Supplier management				■						■
Incident management							■			■
Problem management							■			■
Configuration management						■				■
Change management						■				■
Release management						■				■

CHAPTER 1

Why do metrics matter?

What is a metric?

The term metric is used in this publication for an individual measurement, which may be represented by a single number, a list, table or chart. A group of metrics is used to build a report, usually with supporting text.

Example metrics in this publication are used to illustrate the important features of different types of metrics. This is generally easier to do with charts and supporting explanatory text. For some organizations, tables will be preferred, and for some types of metrics, tables are better than charts.

Although example metrics and reports relevant to each clause in ISO/IEC 20000 are given in Chapter 11, this publication does not attempt to provide suggestions for all metrics within the scope of ISO/IEC 20000, nor should the use of charts be taken to mean that ISO/IEC 20000 requires charts. There is no requirement for the type of metrics used, or how they are presented. What is required is that the service reports that are produced are suitable for informed decision-making and effective communication i.e. the service reports meet the needs of the intended audience.

How are metrics used?

Well-designed and timely metrics are essential for managing, improving and maintaining the efficiency of the management system and service management processes.

With a good set of metrics managers can plan with confidence, deliver a consistent service to the customer and use available resources cost-effectively. Both the service provider and customer can use the information to establish goals for service improvements and other changes and as a basis for critical success factors. Metrics covering topics that are of interest to a customer can reassure them that their business needs are understood and that their service provider is professional and competent.

The principles of best practice service reporting are:

- understand the source of the data used, and the limits to data accuracy that result from the method of data collection;
- ensure that the algorithms used are understood and documented;
- ensure that the design of both the content and format meet the real needs of the target audience;
- periodically review reports to make sure they are still required;
- keep metrics under the control of change management.

ISO/IEC 20000-1, 6.2 requirements

The objective of the service reporting process is to produce agreed, timely, reliable and accurate reports for informed decision-making and effective communication.

The requirements for service reporting are relatively short, but have implications for the whole of service management.

Intelligent use of well-designed and useful metrics is essential to best practice service management, and to achieving ISO/IEC 20000. Auditors will expect to see evidence not only of metrics being produced, but also of how they are designed and used.

Service management process metrics

Service reporting is used to manage service management processes, process interfaces and organizational interfaces. For this reason service reporting is a separate process in ISO/IEC 20000, rather than a part of each process, because it is fundamental to achieving ISO/IEC 20000.

The service reporting requirements in ISO/IEC 20000 include information produced by the service provider for customers and the service provider's own organization, as well as information about the service provided by suppliers via the supplier management process. Where there is a complex supply chain, the service provider should use metrics that reflect the relationships between the service provider and the various suppliers. For example, if several sub-contracted suppliers are managed by a single lead supplier, the lead supplier should report on the whole of the service they provide, including any services provided by sub-contracted suppliers.

The management of the 'supplier ➤ service provider ➤ customer' supply chain is described in detail in BIP 0033, *Managing end-to-end service*.

Metrics for management of interfaces

Simply producing metrics on individual service management processes is not enough to meet the service reporting requirements of ISO/IEC 20000, which includes all measurable aspects of the service and of service management.

The information passing between processes (and across a supply chain that includes different organizations) is included in an audit in order to assess how well the interfaces are understood and managed e.g. information on changes passing to the problem management process (and vice versa) or suppliers' service level reports passing to the service provider via the supplier management process. This is described in greater detail in BIP 0038, *Integrated service management*.

Management responsibilities (ISO/IEC 20000-1, clause 3)

ISO/IEC 20000 also requires metrics on progress with management responsibilities, such as staff competence, awareness and training for all those that are involved in service management.

Metrics and Plan-Do-Check-Act (ISO/IEC 20000-1, clause 4)

As a management system standard, the reporting requirements of ISO/IEC 20000 include the requirements of the Plan-Do-Check-Act (PDCA) cycle. The service reporting process is itself the subject of the PDCA cycle, which checks that the process is efficient and effective, and if not, identifies and acts on improvements to the process, as described in Chapter 7, Documenting metrics.

Implementing a new service (ISO/IEC 20000-1, clause 5)

Metrics used for the PDCA cycle are similar to those used when planning and implementing a new service. When implementing a new or changed service, it is particularly important to plan the metrics, reports and the service reporting process required, both before and after the change.

The link between policy and process

ISO/IEC 20000 requires policies to direct service management processes, including the service reporting process. Whatever the details of a service provider's policy for service reporting, the policy must be supported by the service reporting process and that process in turn must be supported by procedures. This is described in more detail in the next chapter.

Automation of service reporting

Although ISO/IEC 20000 does not specify requirements for automation in order to collect, analyse and report on services, it is recognized that the production of most metrics will not be cost-effective unless there is a level of automation for monitoring, analysis and reporting. Manually produced metrics are also more likely to include errors.

CHAPTER 2

Policy, process and accountability

Policy implementation

ISO/IEC 20000-1, 3.1 includes a requirement that management are responsible for ensuring that there is a policy for service management and a policy for service improvement (ISO/IEC 20000-1, 4.4).
This was included because the the management direction provided by policies form part of the management commitment to best practice service management, which must be demonstrated during an audit.

This encompasses the service reporting process. A policy, formally expressed by senior management describes the overall intentions and direction of a service provider. One of the major benefits of aiming for ISO/IEC 20000 is the establishment of formal service management policies.

Figure 1 shows the relationship between policy, process and procedure that is important to achieve ISO/IEC 20000. It also shows the central role of the service reporting process in service management linking policy, process, procedure, service management plans and service management documents such as service level agreements (SLAs). The relationship between policy, process, procedures, and plans for improvements is also described in BIP 0030, *Management decisions and documentation.*

A policy that cannot be monitored and reported against is bad practice and does not meet the requirements of ISO/IEC 20000. An auditor presented with a policy that had been worded ambiguously or where progress in implementing the policy could not be measured objectively, would be justified in rejecting the policy as unsuitable.

The top-down approach of ISO/IEC 20000 requires processes to support policies, for the whole of service management, as shown in Figure 1.
A process, within a management system, is an activity using resources to transform inputs to outputs. Usually, the output from one process will directly form the input into another process. A good process is one where the activities are timely, effective and repeatable. For service reporting, data is transformed into information.

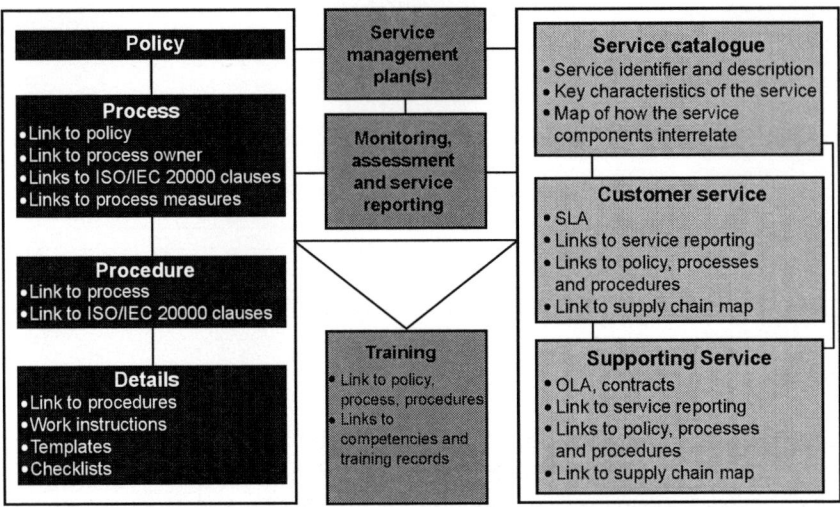

Figure 1 – Relationships in service management

Service reporting policy

The ISO/IEC 20000-2 recommendations on service reporting policy cover the service provider gaining agreement on service reporting for customer and service provider use, covering all measurable aspects of the service, providing both current and historical analysis. The recommendations on policy also cover reports on the services provided by suppliers. This type of report is described in Chapter 5, *Target audience*.

ISO/IEC 20000 requires there to be a clear description of each service report including its identity, purpose, audience and details of the data source. It is advisable for this to be stated by the service reporting policy.

The use of service reports in service improvements is also important, with metrics being used in the PDCA cycle. In turn, the service reporting policies should be supported by processes, procedures and by plans for improvements. The plans for improvements should cover improvements to service reporting process itself, but service reporting is also important to all other processes within the scope of ISO/IEC 20000.

Metrics and policy implementation

Metrics play an important role in reporting progress against policy implementation. This is illustrated for a service improvement policy in the following example.

 Example: Service improvement policy implementation

If a service provider's policy includes the following provision:
'service improvements will be targeted at delivering a faster and cheaper resolution service', then suitable metrics could report on:

- trends in incident and problem volumes
  (cost reduction from problem avoidance);

- trends in average fix time for all incidents and problems
  (cheaper and faster due to quicker methods of resolution);

- trends in unit cost of incident/problem solving, with targets
  (cheaper unit costs).

These metrics are from several processes (in this example, the processes are incident and problem, IT budgeting and accounting).

Lower level metrics will underpin the high-level metrics. For example:

- number and types of incidents and problems in each priority class;

- average fix time for incidents and problems, for each priority class;

- type and number of known errors;

- number of known errors eliminated permanently;

- cost of fixing incidents, including overheads.

These can be underpinned by even lower level metrics that are narrower in scope and often technical, reporting on individual processes or aspects of a procedure. This level of detail is normally produced for diagnostic purposes, rather than as SLA/performance reports. Typical metrics of this type could be:

- fix times for each hour of a working day;

- fix times for each person on a service desk;

- the causes of the most frequently occurring incidents.

Accountability

Senior responsible owner

ISO/IEC 20000 includes a 'senior responsible owner' who is accountable for the overall delivery of the service management plan and that a decision-making group with sufficient authority to define and enforce policy supports the senior responsible owner.

The service reporting process owner should work very closely with the senior responsible owner because of the extent of service reporting and its interface to every process within the scope of ISO/IEC 20000.

Process ownership

ISO/IEC 20000-1 includes a requirement that each service management process, is owned by a manager who is accountable for its quality. The process owner must be an active champion of the quality of the service reporting process and have the seniority, and sufficient management backing, to influence the effective operation of the process. The process owner must ensure that the service reporting process is properly integrated with other processes (often across organizational boundaries). This is described in more detail in BIP 0038, *Integrated service management.*

The service reporting process owner should ensure that:

- the service provider's policy on service reporting is appropriate for the business needs of the service provider and customer;
- the process is appropriate to the delivery of the policy;
- the service reporting requirements are documented;
- the process is mapped out and documented in procedures;
- there is suitable automation of the production of metrics taking into account the range and complexity of the data involved;
- reports are maintained, updated and improved to meet the needs of the target audience.

CHAPTER 3

What is a good report?

A good report is normally one that includes metrics that illustrate the speed, effectiveness and predictability of a process and, therefore, how well a policy is implemented. A good report is also timely, clear, reliable, concise and appropriate to the recipient's needs. It is sufficiently accurate to be used as a decision-making support tool. Good reports are designed so that the information they contain is easy to assimilate and allows the reader to identify what action should be taken as a consequence.

Reporting objectives

The reports should include metrics that focus on those things that matter and those that can be changed by the service provider.

For example, to:

- monitor and report actual service levels against agreed targets or objectives;
- understand the workload and how and why it changes;
- provide reports that include supplier performance;
- select underpinning metrics to support:
  a) management of non-compliance/escalation;
  b) identification of service trends;
  c) preventative actions or procedures.

Metrics should also enable the reader to determine, objectively, whether the service is:

- cost-effective;
- meeting agreed targets;
- within predicted workload levels;
- responsive to the customer's changing business needs.

The same metrics may be of interest to both the customer and the service provider, although possibly for different reasons. It is the combination of the report type and target audience that determines the purpose of a report.

Example reports

Under the PDCA cycle and the service reporting process, a service provider typically produces reports for customers and for internal service management covering:

- performance against service level targets and service level agreements (SLAs);
- non-compliance and other issues (e.g. security breaches, outages);
- workload characteristics and volume information (e.g. numbers of incidents, problems, changes and tasks);
- performance reporting following major events (e.g. major incidents and changes);
- trend information with the reason for a trend explained (e.g. calculated on an hourly, daily, weekly or monthly basis);
- future and scheduled workloads, including extrapolation of historic data to give predicted values;
- satisfaction analysis and other customer views;
- management system responsibilities such as headcount, headcount by type of role, turnover rates, length of service in grade, and training and skills at each grade.

Reports may be specific to a single process, such as the number of incidents, the most frequently asked questions, the unreliable components of the infrastructure, and the most resource intensive tasks. Others span processes and may be used in the management of the interface between processes e.g. customer satisfaction reports feeding into service level management and business relationship management. Other examples of interfaces include incident data feeding the problem management processes (and vice versa), and capacity management metrics feeding the continuity process.

Additional information on metrics is given in the other publications in the 'Achieving ISO/IEC 20000' series, and in the publications listed in Appendix B. Example metrics are also given in Chapter 11.

CHAPTER 4

Types of metric

Having described metrics and reports and the range of information they can include, this chapter identifies the types of metric available and those most commonly used in producing reports. Although it is not mandatory for a service provider to do so, ISO/IEC 20000-2 subdivides reports (and individual metrics) into categories based on the three main types of information they provide. These metrics are described in this chapter, and are:

- reactive;
- proactive;
- forward schedule.

Reactive reports

Reactive metrics show what has happened, particularly from the predominantly reactive processes, such as incident management. Reactive reports are generally the first type of report to be produced when service management processes are implemented. They are normally the easiest to produce and can be very important to the management of the service and to the customer. There are many different types of reactive metric, but examples include:

- service desk call volumes per month;
- problems and incidents (ordered by type, impact or urgency);
- security incidents;
- number of changes;
- response times.

Example: A reactive metric

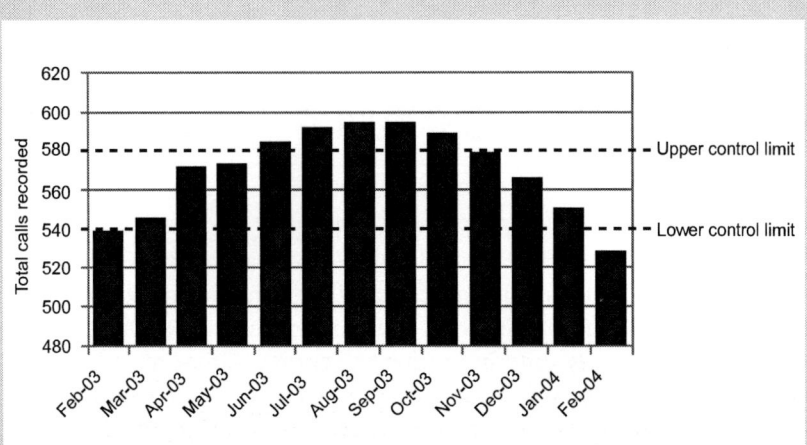

Figure 2 – Call centre example

Call volumes per month

The number of calls actually recorded every month can be tracked to provide reliable information showing how call levels have changed over time. This is commonly one of the first metrics produced by a service provider.

From a metric of this type a customer could gain an understanding of what the service provider was dealing with on their behalf in very simple, high-level terms.

A service provider could use this information in headcount-planning but should also seek an understanding of the reasons for any trends or unusual features. For example, the workloads may have peaked during the middle of the reporting period. This may be seasonal or it may be related to a specific event that will not recur. If the period is extended beyond a year, the impact of seasonal trends may be identified and better understood for workload management.

Understanding the cause of the peak in workloads is important for planning the management of future workloads, and for any changes that will prevent the need for calls i.e. proactive use of a reactive report for call avoidance.

The service provider could also establish upper and lower control limits, as shown by the dotted lines in Figure 2, so that a review is triggered if the value falls outside the upper and lower control limits. A service provider may also choose to only produce the metric when the values for that month fall outside the limits of normal workloads represented by the upper and lower control limits.

Proactive reports

Proactive metrics give advance warning of events that if left unmanaged could impact the service. If acted upon within the right timescale, they will enable preventive action to be identified and taken before the service has been impacted. This type of report is frequently based on the extrapolation of historic information. Proactive metrics may be predictions based on the estimated impact of expected events, such as workload increases coinciding with a new product launch. Examples of proactive reports are given in Figures 3 to 5.

Example: A proactive metric

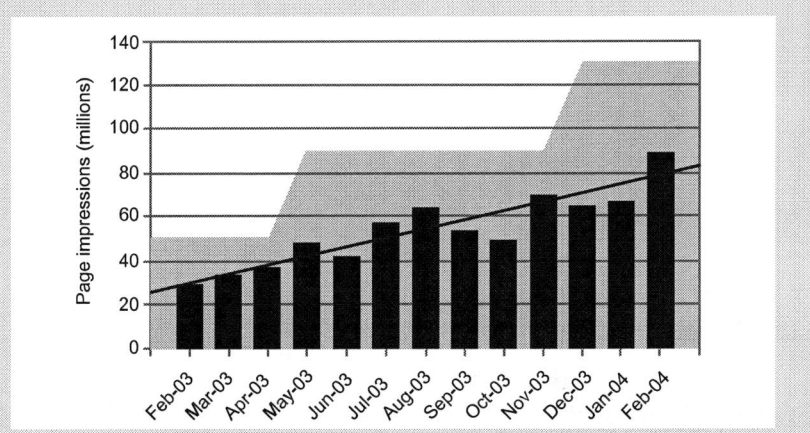

Figure 3 – Trend in the number of web page impressions

Web page impressions

This metric is a chart that reports the number of web page impressions as columns, with a line chart showing the trend in impressions over a thirteen-month period. The actual values are compared to the maximum number of web page impressions predicted for the available capacity. The capacity has been increased twice during the thirteen months that have been reported.

A customer could use this report to assess the changes in popularity of their web-based service during the thirteen months. More detailed information on the activity generated by the new web pages could then be gained from examining more detailed, lower level information.

The service provider might use this information to make capacity projections based on a combination of the trend information and their understanding of the customer's business plans for their web-based service.

Example: A proactive metric

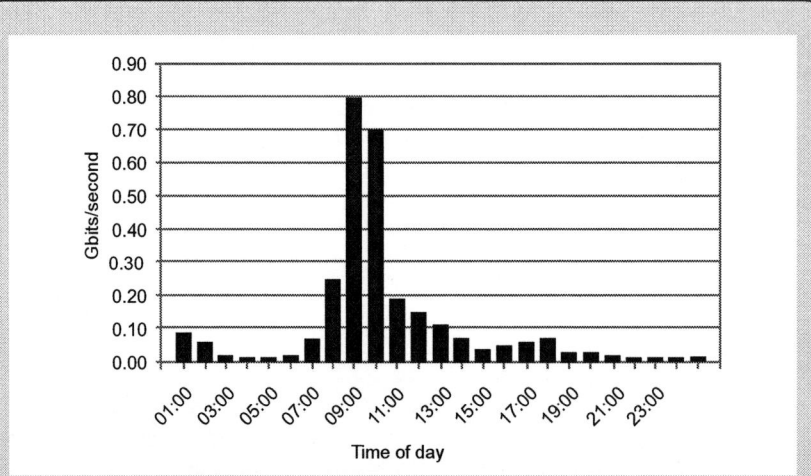

Figure 4 – Day workload profile

Workload profile

This metric shows trends in monthly averages and is more detailed than Figure 3. It shows how the workload changes during a 24-hour period. This metric could represent a single day, or be an average for several days.

This chart shows that a large peak in activity occurs early to mid-morning, with little activity outside this peak period. Capacity that is adequate for the average activity level would be inadequate for the peak activity.

A decision would have to be made regarding the capacity, and whether it should accommodate the maximum activity represented by the peak, or the average activity, or compromise between the two, such as a percentage of the peak.

Forward schedule reports

Forward schedule reports show planned activities, i.e. those activities that are not just predictable but which are intentional. Most forward schedule reports are part of a change lifecycle and many are the result of individual projects, e.g. the changes needed for the implementation of new services or the launch of new products. In addition to documents such as Gantt charts, showing planned events, the following are examples of forward schedule reports.

Change management reports

These alert the reader to a planned change, clearly identifying any potential risks to the customer's service and the timing of any planned service outages.

This type of metric will often be text based to allow for the level of detail that is required of such a report. However, a graphic can be used to illustrate events on a calendar or timeline.

Capacity management reports

These show projected capacity requirements based on scheduled events. They can be used to show planned changes to capacity that is required when the current capacity will become inadequate.

This planning will avoid an unacceptable impact on the service from inadequate capacity. Normally the report will also explain what the service provider is planning to do to avoid impacting the customer's service.

Capacity issues can manifest themselves in a variety of ways, e.g. increasing response times or business activities being held up by a procedural bottleneck.

In Figure 5, the capacity in question is the maximum number of security tokens that are available for a remote access service, showing the increase in tokens planned.

Example: A forward schedule metric

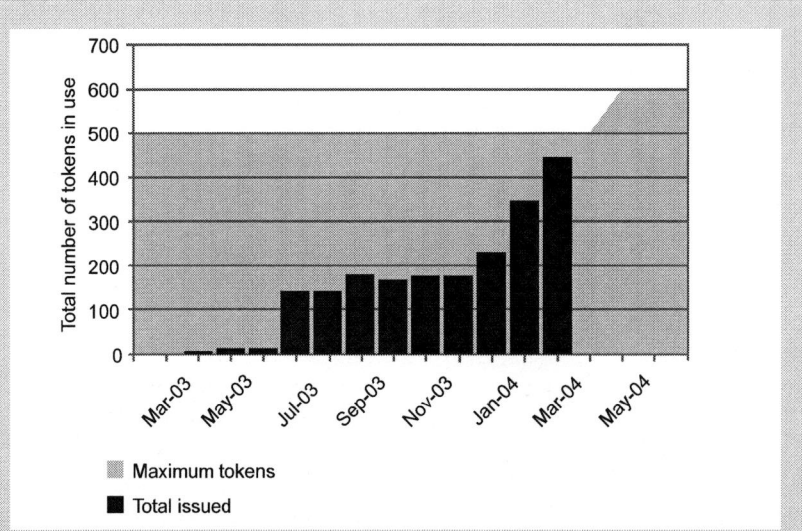

Figure 5 – Security token usage and planned increases

Security tokens

This metric tracks the number of security tokens available for a remote access service and those that will be required in the future. A trend line could be added to project the actual number of tokens in use, to demonstrate when the currently available tokens will have been allocated. This type of forward planning report can also be used as part of a business case that is justifying funds for another batch of tokens before the tokens have actually run out. It can also be used to illustrate the need for security controls around the tokens, such as the need for:

- unused tokens to be held securely;

- unused tokens to be retrieved;

- the growth of remote support to be capped (especially if no budget has been allocated for more tokens to be bought beyond the initial 5,000 tokens).

CHAPTER 5

Target audience

'The target audience for a metric is also a method of categorizing a metric. For metrics 'one size does not fit all', and reports should be designed for a specific target audience and not used indiscriminately for everyone. The selection of metrics that are of interest may also change over time, as changes are made to the service or there is a change in the business activities of the customer.

Who is the target audience?

The target audience in terms of ISO/IEC 20000 activity is generally going to be the customer and the service provider.

Customer reports describe the service from the perspective of the customer, often linked to an SLA. See *Customer reports* later in this chapter.

The service provider's own metrics should cover each aspect of service management but should also report on the status of the service provider's management responsibilities, defined in the requirements in ISO/IEC 20000-1, clause 3. The metrics that are required for the PDCA cycle defined in clause 4 of ISO/IEC 20000 are also part of service reporting. These are described at the end of this chapter.

Under some circumstances the same metric may be used by both customer and service provider, but with different objectives. For example, a report may be logged by support staff on the number of 'How do I...?' queries, and may be used by:

- the customers to identify future training needs;
- the service provider and the customer for the development of 'Frequently Asked Questions' (FAQs) to help end users; and
- the service provider to assess the effectiveness of the release process under PDCA.

Depending on circumstances, each of the types of report used in ISO/IEC 20000 i.e. reactive, proactive and forward schedule, may be suitable for either the customer or the service provider (see Chapter 4, Types of metric).

Customer reports

The main groups of customers are:

- senior customers;
- stakeholders;
- end users.

It is not normally productive for a customer to request a very large number of metrics 'just in case', as the cost of production may only have the effect of making the service more expensive. Often, a cost-effective approach for the service provider can be to produce a set of core metrics that are relevant to all customers, with additional metrics that can be used for specific customer interests, instead of automatically applying all metrics to all customers.

Customer terminology

Metrics that report on the service delivered to customers and users have an external business focus and should use the customer's terms and language, not the technical language or jargon that only the service provider uses. Customers should also have metrics that, where possible, state the business impact/cost rather than, for example, showing trends in urgent problems that have been fixed by each support team. These customer reports will typically include information such as:

- problems with system x, delayed billing for y days, with loss of z;
- power loss resulted in end users unable to work for x hours with costs to the business of y.

Purpose of the reports

External, customer-targeted reports on performance should help the reader answer the following major service related questions.

- Is the service delivering what the customer expects?
- Is the service delivery meeting the customer's business needs?
- Is the service value for money?
- Has the service got better or worse?
- What can be done to improve what the customer sees?
- What can be done to improve the service delivered to the customer?

Typical examples of customer metrics and reports include metrics for SLAs, exception reports, and cost/budget reports, as illustrated in Figure 6. Others will be required, as described in BIP 0033, *Managing end-to-end service*.

Metrics for service level agreements (SLAs)

Any report that describes services included in an SLA is by definition a customer's service report. These usually compare actual service levels to service level targets.

SLAs vary widely across service providers depending on the type of service and the customer's business needs. There is no single set of targets that applies to all service providers, and therefore no single set of metrics is suited to all customers.

An example SLA metric is given in Figure 6. Metrics of this type may have textual explanations describing the reasons for the changes to the service levels shown.

Exception reports

An exception report is an appropriate method of understanding and explaining an unusual event that has either impacted the customer's service or put it at risk, such as a major problem, a serious security incident or an unexpected change in workload volume or workload characteristics. Many of these reports are predominantly text based as they require detailed explanations of events and not just a metric relating to the event.

Exception reports also reduce the number of reports routinely issued which actually report nothing of interest to the target audience. Suppression of a routine report could be by manual intervention or by the reporting system, based on rules, so that it is produced only under exceptional circumstances. The definition of exceptional circumstances should be agreed by the target audience. The service reporting process owner (the role of process owner is described in Chapter 2, Policy, process and accountability) should be involved, as well as the owner of the service level management process or the owner of the business relationship management process.

Example: Customer metric 1

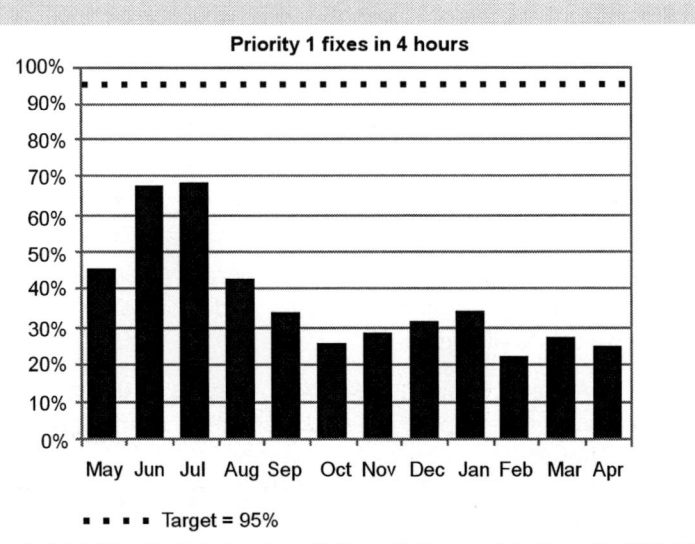

Figure 6 – SLA metric for a customer

Priority 1 fix times

This is a commonly used type of metric, which illustrates the actual percentage fix time compared to an SLA target of '95% fixed in 4 hours'. This chart shows that the SLA target has not been achieved during the period reported, and it also shows that there has been a gradual degradation in the service, shown by the downward trend in the percentage fixed in the target time.

A customer who is reliant on Priority 1 fixes being carried out quickly to keep their business operational would be very concerned by the failure to meet this service level target. This level of service could have a serious impact on the customer's ability to function as a business. A competent service provider would have promptly investigated the breach in service levels, established why it was occurring, and what could be done to ensure that the targets were being met, or at the very least to stop the continuing degradation.

An auditor would wish to see evidence that this service level had been reviewed and consideration given to how the below target service should be addressed. The starting point for this is to understand why the degradation had occurred. The following are examples of some of the many reasons for this.

- Has the number of Priority 1 fixes increased over the period reported?

(This is masked by the use of a percentage rather than the actual number.)

- Has the nature of a Priority 1 fix changed over time?
- Are there more complex problems that are difficult to fix arising from a change to the service?
- Are other activities distracting the service provider's staff from working on Priority 1 fixes?
- Are less staff allocated to deal with Priority 1 fixes?

Lower levels of detail may be used during the review e.g. the number of Priority 1s, the types of Priority 1s and a report matching the Priority 1 workload to the staff available.

Not all reviews will lead to change e.g. a review of the underlying reasons may identify that improving the service level requires an investment that has either not been budgeted for or is considered to be too large to make the cost of improving the service levels worthwhile. The auditor would expect to see that the results of the review had been discussed with the customer and that the action (or inaction) had been agreed.

Cost or budget information

Even if a service is not charged for, most customers will have an interest in the cost (and perceived value for money) of a service, but particular attention will be paid to a service that is thought to be expensive or does not provide the appropriate quality of service required. Financial reports usually consist of cost and budget information in tabular form, with some supporting charts and text.

Cost or budget information combined with service information is more likely to be of interest than two separate reports. For example, support costs per PC may be preferred to a separate report on the number of PCs and a separate report on the cost of supporting PCs. Similarly, unit costs of an incident or change, costs of email storage/email accounts, and unit costs of 1,000 web page hits may also be preferred to separate reports. This is particularly the case when budgets are being prepared or when the customer has had their own budget reduced. This may also be the case if the customer intends to cut the service provider's budget or is asking the service provider for improved cost-effectiveness. More details are provided in BIP 0034, *Finance for service managers*. Examples are given in Figures 7 and 8.

Example: Customer metric 2

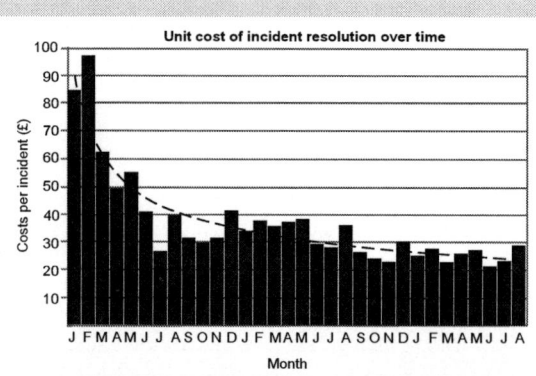

Figure 7 – Example cost metric

Unit costs of incident/problem fixing

This type of metric is particularly useful for senior management, who are interested in costs as well as service quality. The illustration refers to 'incidents'. In ISO/IEC 20000, the terms incident and problem are aligned with the terms used in the IT Infrastructure Library (ITIL®). During the metric design stage, what is to be reported must be established i.e. problem, incident, major incident, known error, as this decision affects the data records included or excluded when the metric is produced.

ISO/IEC 20000 uses the terms 'incidents' and 'problems' for two different events, a difference that will not normally be understood by customers. A service provider may choose to group incidents and problems for a customer report, but report on them separately for their own internal reports. Other specialist ISO/IEC 20000 terms will also be unknown to the customers and may also be unknown to some of the service provider's staff, emphasizing the need to understand what is reported by each metric.

The cost basis must be defined. Typically this includes:

*   all support staff costs relevant to this type of work;

*   staff overheads (a finance department will be able to give an indication of this using the organization's normal financial rules for what is an overhead and how much should be allowed as an addition to basic staff costs);

*   other overheads, such as office space and service management tools;

*   the costs included or excluded e.g. software development costs are normally excluded, but software maintenance costs are normally included.

Unit costs are one total divided by another. The chart shown in Figure 7 is useful for tracking efficiency improvements.

Example: Customer metric 3

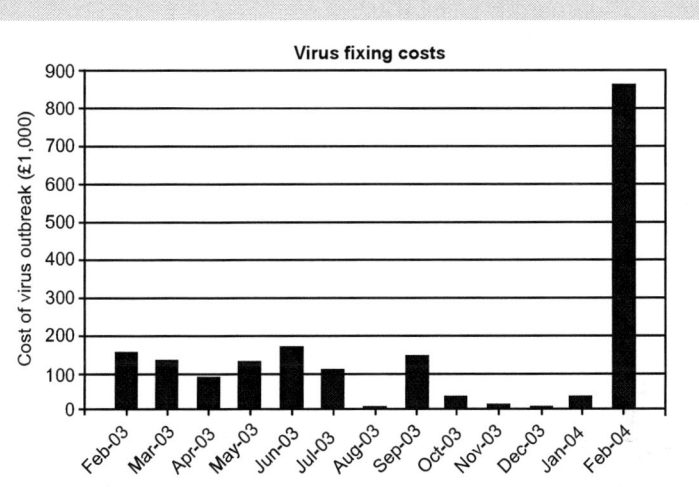

Figure 8 – Example cost metric

Costs of virus fixing

Rather than report on the number of viruses that are fixed, a metric could be produced to show the cost of fixing viruses based on the cost of the support staff required for this. This metric could be extended to include the cost of end users that are unable to work (lost business). The combination of cost, workload and number of viruses in a metric will be of particular interest to senior customer management, as well as process owners.

The basis of the cost calculation could be as simple as using the time between the virus being reported and the time it was fixed, and applying standard hourly rates. If the information is given to the customer, the basis of the calculation should be agreed with the customer. The values shown in Figure 8 are for illustrative purposes only.

The cost of virus fixing could be used to explain the need for a rigorously applied security policy designed to protect the infrastructure from viruses, but which may be unpopular with end users and therefore regularly flouted. The impact on the customer's business of a new virus may be huge, and may be much greater than the cost incurred by the service provider for actually fixing the incident. Business time that is lost is a useful incentive for ensuring a security policy is actually adhered to.

Customers may also wish to have access to the service provider's metrics to help them manage their dependency on the service provider. The service provider reports could be of interest to the customer for a variety of reasons, including the following.

- If the number of calls made to the service desk and handled by the incident management process are increasing then the customer may want to understand the underlying cause of any changes or trends.
- The customer may want to know which questions are most frequently asked of the service desk.
- The reports on the cause of serious service failures, used by the service provider for preventing recurrence, may also be of interest to customers.
- Metrics on the types of events handled by the service desk can aid decision making on whether further training, education or user guides are needed.
- The customer may want to assess whether changes (e.g. in data validation) would be effective in reducing support demand and the cost of the service, whilst at the same time increasing user productivity.

Service provider reports

For the service provider, the main groups are:

- business relationship managers;
- service provider senior managers;
- process owners;
- managers and staff responsible for delivery of the service on a day-to-day basis;
- supplier managers.

Service provider terminology

The terminology used and the overall view of the service will be from the service provider's perspective and may be much more technical than the reports produced for customers. Service management terminology will be meaningful to any service provider that is aiming for ISO/IEC 20000, e.g. the term 'configuration item' or 'release'. Other examples are reports on network failures, security issues or infrastructure bottlenecks that result in performance problems.

Purpose of the service provider's metrics

The service provider's own metrics are internally focused and are usually much larger in number than those produced for the customer. They cover the identification of underlying faults and workload/performance data. They should reflect the components of the service, e.g. components such as servers, routers, hubs and cabling.

Service managers should also have reports to help them manage the people, processes and the interfaces between processes and different organizations. The reports should help to identify:

* service trends;
* unreliable components of the infrastructure;
* resource/cost intensive tasks.

Typical examples

Service provider reports normally include metrics that are scoped by technology, a support team or process, and not by a customer's business unit. These are commonly based on:

* components of the service, e.g. server availability;
* the performance of a functional group in the service provider's organization, e.g. problem fix rates of a second-line support group;
* process, e.g. change management.

Service provider metrics may include a low level of detail.

They may be dynamic, with the metrics being generated and used on a continuous basis rather than as part of a daily, weekly or monthly pattern. For example, fault monitoring may be tracked on a screen, with alerts set to trigger when defined control limits are exceeded. An auditor may wish to see this type of metric demonstrated on-screen, and also evidence of this type of information being used for management of the service, and where appropriate as input to the PDCA cycle. The parameters that determine the alerts need to be documented and under change control.

The following are examples of internal service provider reports.

Reliability of a network component
This report is used for the management of the network. This type of report is normally very important to a service provider for diagnostic purposes, enabling the identification of improvements.

Capacity: actual and available

Tracking the use of a resource and its available capacity can give the service provider advance notice of the need to increase capacity. Capacity management reports are described in BIP 0037, *Capacity management*.

Trends in the number of incidents, by type

Monitoring the trends in incidents can provide useful information to the internal management processes of the service provider. For example, a change over time in the proportion of incidents in each priority category should trigger an investigation.

This may establish that:

- the method of allocating priorities is changing;
- the customer's business needs have changed;
- certain types of incident are occurring more often.

Incidents may also vary across different locations, types of user and technology.

An auditor will seek evidence of how information is used in service management. For example, the service provider can make a decision to do one or more of the following.

- Invoke escalation procedures if a major incident has occurred, or if an incident was unresolved well after the fix time target.
- Reallocate resources to best meet the customer's needs, taking into account the nature of any competing priorities, the nature of the work that needs to be done, and any time critical issues.
- Re-educate the support staff in how to set and agree priorities with customers.
- Improve user awareness and education if there are many incidents or requests of the same type. Frequently asked questions can also be developed and issued.

Service provider metrics provided by suppliers

A service provider has responsibility for the whole service that is delivered to the customer, and therefore needs metrics that cover the whole service. The service provider should include metrics and reporting needs when specifying the service they require from their suppliers. Reports provided by suppliers are of interest to several processes, not only the supplier management process.

When there are multiple suppliers, the metrics are normally provided by each of the suppliers under the terms of the contract (or other formal agreement) between the service provider and each of the suppliers.

The metrics should reflect the relationships between each organization, i.e. the service provider, suppliers[1], lead supplier and sub-contracted suppliers. For example, if a lead supplier is in place, the lead supplier should provide a consolidated view of the service provided by the lead supplier and each of their sub-contracted suppliers. Alternatively, if there are multiple suppliers, and each is managed directly by the service provider, each supplier should provide metrics that report on their delivery against the commitments defined in their contract. In this case the service provider consolidates the service reports from each supplier.

As each organization involved in the supply chain will have to report on the service they provide, it may be beneficial to integrate reporting systems to ensure that the right data is flowing between organizations, or possibly to establish a single system. However, this is not a requirement of ISO/IEC 20000-1.

Similarly, service providers will find it beneficial if their suppliers adhere to the same design principles when developing reports. This will ensure that the supplier's metrics and reports, which are required to manage the interfaces between processes (and the whole supply chain), are compatible with those of the service provider.

Metrics on management responsibilities, ISO/IEC 20000-1, clause 3

Some of the requirements of ISO/IEC 20000 are related to management responsibilities and evidence of this will be sought during the ISO/IEC 20000 audit. They include a wide variety of possible metrics, such as:

- progress in establishing the service management policies, objectives and plans;
- progress with communicating objectives and the importance of ongoing improvements;
- progress of the service management plan;
- status and plans for service management documentation, including that which is required as evidence during a ISO/IEC 20000 audit;
- progress in establishing full process ownership;
- progress in documenting procedures;
- progress with the definition of roles, responsibilities and competence assessments;

---

[1] In ISO/IEC 20000 the terms 'lead supplier' and 'subcontracted supplier' have specific meanings used to describe their place in a supply chain. See BIP 0033, *Managing end-to-end service*.

- status of risk management;
- assessment of training needs;
- actual training records against planned and budgeted targets;
- progress in staff development to meet new or changed needs.

 Example: Service provider metric

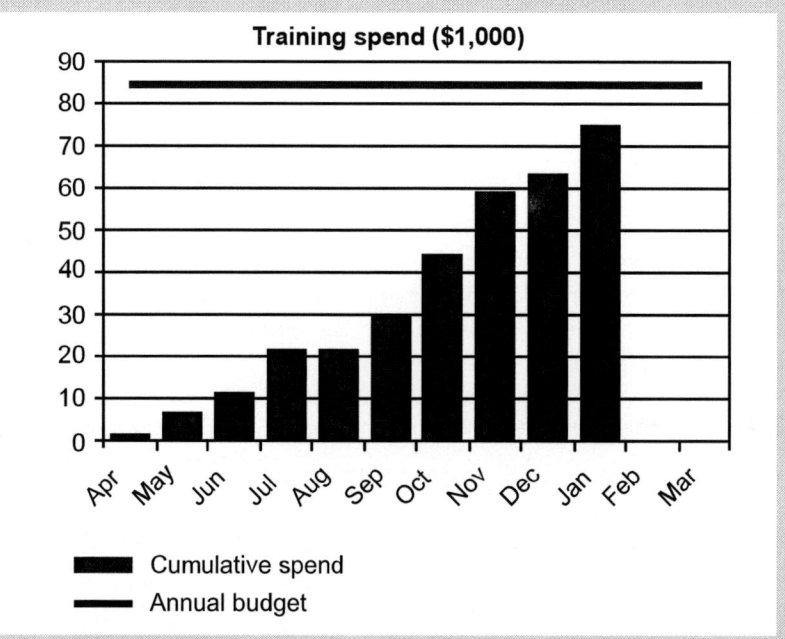

Figure 9 – Management responsibility metric

Staff training spend

ISO/IEC 20000 specifies requirements related to the competence, awareness and training of staff, with the objective of ensuring they have appropriate skills to meet the customer's needs. A service provider may report on various aspects of this, e.g. the total amount of money spent on training staff per month.

A report like the example given in Figure 9 shows the cumulative spend and the annual budget ($84,000) enabling the reader to calculate the money available within the rest of the financial year. This can be either in the form of a chart or table (or a combination of both) with supporting text.

Other aspects that could be reported on include training by type, the availability of skills based on the levels of competence reached.

Metrics for PDCA, ISO/IEC 20000-1, clause 4

The PDCA requirements are also included in the scope of the reporting process. Typical questions that may be asked as part of the PDCA cycle include the following.

- Is the process supporting the relevant policy[2]?
- Is the process supporting the service management plans?
- Is the process delivering its objectives?
- Is the process efficient?
- Is the process value for money?
- Has the process got better or worse?
- What can be done to improve the process/service management?
- Have objectives for process improvement been identified (often referred to as 'stretch targets')?
- Are the interfaces between processes managed?
- Are organizational interfaces being managed?
- Is there demonstrable continual improvement?

Metrics linked to the PDCA requirements are primarily of interest to the internal service provider. Customers are most likely to be interested if there have been, or are about to be, major changes such as service improvements.

The relationship between service reporting and PDCA is described in more detail in Chapter 8, PDCA cycle and service reporting.

---

[2] For the relationship between policy, process and procedure see BIP 0030, *Management decisions and documentation*.

Example: Service provider metric

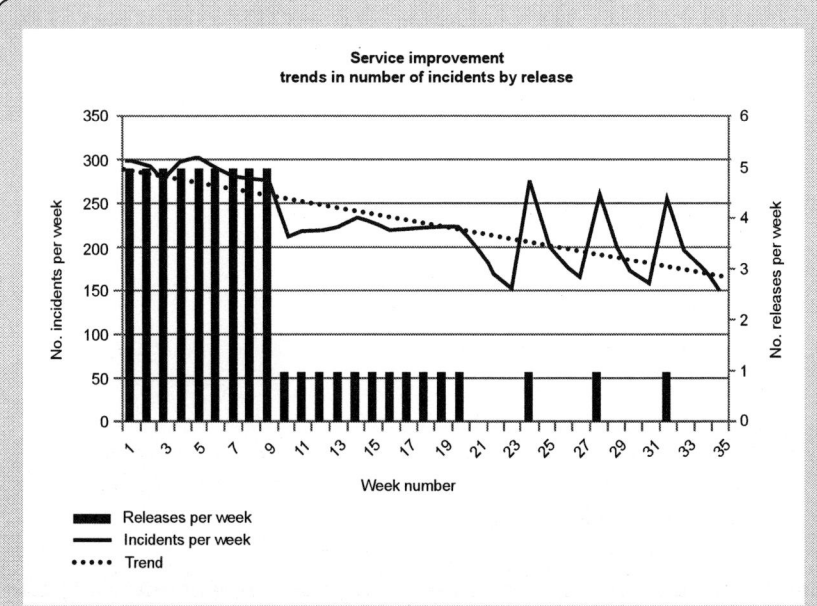

Figure 10 – PDCA / release management metric

PDCA and the release management process

Metrics could include checks on the effectiveness of the release of new software to the customer via the service provider.

Metrics could report the number of incidents caused by release management, per week (shown by the bar chart in Figure 10). In this example the release policy was changed at week nine so that several small releases were batched to schedule one larger release per week.

The chart shows that the number of incidents that were directly attributable to the release actually reduced (shown by the line chart in Figure 10). After a further review the release policy was changed again so that there was only one release per month. Although the number of incidents per week reduced further, they were variable and harder for the support team to manage. As the general trend was a reduction in the number of incidents caused by releases, the release policy remained unchanged as the release testing was improving and the time spent managing the release was reducing.

The efficiency of the process could be checked before and after the changes to the process, and made to meet the needs of the revised policy.

CHAPTER 6

Design guidelines

Top-down versus bottom-up

It is common, but undesirable, for the designers of customer reports to start with metrics that are detailed, i.e. technical reports that are produced by the service provider for internal use, which are then used as reports for customers. This is because they are already available and therefore cheap and easy to produce for the customer. This is a bottom-up approach to the design of metrics.

A bottom-up approach usually generates too many metrics, which will mask the important metrics in a mass of irrelevant detail. Producing a large number of metrics when only a small subset is required is a common failing of service management, not a strength as it is sometimes perceived to be.

A top-down approach is preferable for the design of metrics. It is generally more productive for metrics to be designed as a hierarchy of information starting with the highest level first. A top-down approach is described in Chapter 10, *Techniques*.

 Example: Service management policy and SLAs

If the service provider's policy includes the following: '*...services shall be delivered as described in an authorized SLA...*' then suitable metrics include the percentage of SLAs that have been authorized out of the total authorized and unauthorized as well as the reports that show and explain service breaches. This could be underpinned by reports on the progress towards SLA authorization, for example number:

- drafted;
- under review;
- being revised;
- rejected;
- awaiting authorization.

A metric hierarchy

Top-down planning results in metrics that have a logical relationship to each other, avoiding gaps, overlaps and duplicated effort. The highest level should be a small number of the most important metrics, reporting against the service provider's vision for services as described in service management policies. These policies and the metrics should also reflect the customer's business needs. The highest-level metric is information based on aggregated data used in lower level detailed metrics.

The very lowest level metric is a building block of data, such as the detail of a single incident or the performance of a single component of the infrastructure. This principle is illustrated in Figure 11. This example shows the aggregation of individual incident and problem/known error data, as well as the aggregation of data on the costs at a detailed level, to provide a high level measure of the unit cost of defect resolution.

The intermediate metrics may be aggregated according to the interests of the target audience, such as incidents affecting a single location, incidents of a particular priority or of a particular type of hardware or software. Metrics may also be aggregated by customer business unit, service or system.

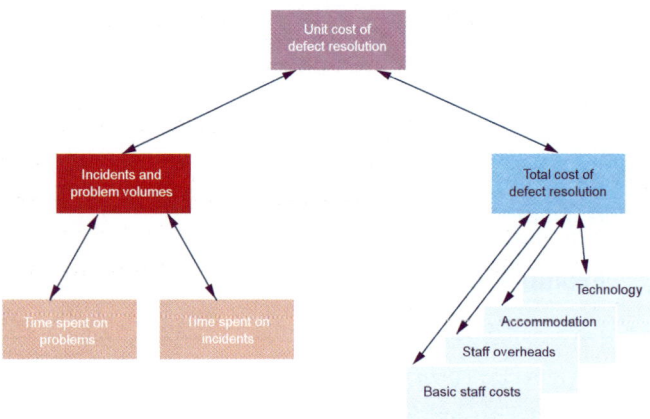

Figure 11 – Hierarchical relationship between metrics

Accuracy of data

Aggregating data also aggregates errors, so during the top-down design of metrics it is important to understand the errors of each component before relying on any of the metrics. Not only will individual errors be compounded as the data is aggregated, but also the high-level metrics are likely to be used for more strategic decision-making. This could have very serious consequences.

High-level metrics are often those intended for the customer, and a disparity between the customer's experience and the messages given by the metrics arising from undetected data errors will be detrimental to the relationship between the customer and the service provider.

An iterative approach

In practice, there normally has to be a compromise made in the top-down design process if it becomes apparent that the low level data is either not available, or too expensive to collect, for aggregation to a high-level metric. This approach still offers advantages over bottom-up design which is normally associated with gaps and overlaps in metrics.

Overall control

Metrics that evolve, rather than those that are planned as part of a larger set of metrics, often overlap. This parochial approach is a waste of effort for the service provider, increasing costs without adding benefit. It is also common to find that similar metrics are based on different algorithms. For example, 'percentage availability' could be calculated by one group based on a 24-hour day, by a second group using a 10-hour service day, by a third group excluding scheduled down-time as lost availability, and by a fourth group including scheduled down-time as lost availability.

Although the service provider's reports are often produced by the people that use them, it is still advisable for the service reporting process owner to have influence over metrics and reports for internal and localized use so that a coherent and logical set of metrics are available for service management processes. An uncoordinated approach will also demonstrate to the auditor that there has been too little management involvement in decisions on metrics.

What metrics are of interest?

It is important to make the link between a metric and the reader's interests as direct as possible. Each group of readers may have different interests and, as a result, will need different metrics. Some groups may have a role that means they do not need metrics at all and may have no interest in the reports they are already being sent. The following are examples of metrics that may be of interest to different groups.

- **Customer satisfaction**
  If there is an interest in customer satisfaction, the service provider should report on this and not the number of customer complaints.

- **Lost service at critical times**
  If a customer is mainly interested in lost service on the last working day of each financial month, it is important to report all incidents during that critical period, with less detailed information on the rest of the month.

- **End user view**
  Show the customer an end user metric that matches their experience, such as how long it takes to fix an incident that affected their business area and not the performance of a component that they may not understand or be aware of.

ISO/IEC 20000 does not require large numbers of reports, but does require that the service provider can explain why each target audience has an interest in what is being reported to them. Brainstorming and workshops are effective methods for generating ideas about which metrics are both useful and possible. A more robust set of metrics and reports will be generated as a consequence. The business relationship management process should also be able to provide input on what will be of interest to customers.

Establishing the interests of the target audience is usually a three-stage process and the stages may have to be reiterated.

- **Explore ideas:** by open discussions of what each group is interested in, particularly customers. The early stages of this process may only cover what is not liked about existing reports, as the first step to understanding how to improve them.

- **Feasibility checks:** to ensure that the suggested metrics are actually possible. For example, the raw data must be available, fit for purpose and cost-effective to produce, as described in the next section, *Presenting the reports*.

- **Provide options:** so that the target audience is able to select the most salient metrics. By this stage it should be possible to broadly predict what metrics are of interest, but presenting a small number of options will help confirm this decision.

Best practice service catalogues, SLAs, operational/internal service agreements, contracts, escalation maps, organization charts and processes details are all linked to metrics, and reviewing one is normally linked to a review of the others.

Once each target audience's interests have been established it should be reviewed at appropriate intervals, as part of the PDCA requirements described in Chapter 8.

Presenting the reports

During the design stage it is important to clarify how the metrics and reports will be made available, i.e. will they be hard copy, emailed or made available on a web page. How and when the information will be used, and what media it will be made available in will be influenced by the interests of the target audience. There is no single correct approach and usually a service provider will adopt a mixture of methods. For example, some internal service provider reports are only required to monitor certain events, and need not be printed or made widely available. Some reports are made available on line, usually if they are mainly of interest to a wide audience. In contrast, reports for customers attending a service review will normally have to be printed as hard copies as they are normally used in meetings.

An auditor will expect to see evidence that the methods used to make metrics available are appropriate for the target audience, and evidence of how the metrics are used, within any technical or cost constraints.

Checklist for metric design

The following checklist summarizes the key points that should be considered when designing new metrics and reports.

Checklist 1: Metric design
Who should receive a report and why?
What does the target audience want to know and why?
In what format would the target audience prefer to receive information?
Will the frequency of the report be appropriate?
Is the cost of the metric production justified?
Are supporting metrics available if required?
Is it clear what process(es) underpins each metric?
Is the relationship to other metrics understood?
Is the data and the data analysis accurate enough for the metric's purpose?
Are the metrics reported to greater precision than the accuracy justifies?
Does the report make the link between service management objectives and critical success factors clear?
What action will be triggered by this metric?

Checklist 1: Metric design (*continued*)

Does this metric report on something where no action is possible?

Who interprets the metric and will they be able to do this accurately?

Is the interpretation objective?

Will the information be in a form that is understandable to the recipient?

How will the metric and service reports be tested?

Essential steps

Do you have the data for this metric?

It is important for a service provider to understand the limits to the metrics it is capable of producing. Automation of monitoring and reporting may not be possible for all metrics. For some metrics, only expensive and potentially error-prone manual data collection methods are possible.

Is this metric useful?

A service provider should only produce metrics that are important or they may distract attention from more important topics. Useful metrics should also be linked to at least one of the following: policy, process and procedure and also to a service or service catalogue entry.

Positive cost-benefit

Cost of data collection and report production (both existing reporting and customized reporting) should be balanced against the benefit of the information gained. Many metrics can be produced with 'fit for purpose' accuracy at a low cost, but with high precision at a disproportionately higher cost. High precision is only cost-effective if low precision might trigger a completely different set of actions. For example, collecting customer satisfaction data to a precision of 0.1% is far more expensive than collecting data to a precision of ± 2%, but the difference in precision will make little or no difference to decisions on how to improve customer satisfaction.

An auditor will only expect to see cost-effective, fit for purpose metrics, and will expect the service provider to demonstrate an understanding of what the potential issues are, and which of those are important and which are not.

What action should this metric prompt?

Metrics are not only intended to inform, but also to help identify and prompt action, encourage the right behaviour and to trigger both process and technological changes. To enable a decision-making process for improvements and other actions, metrics must be easy to understand so that there is clarity on the information provided. A metric that cannot be used to prompt action is not normally worth producing, e.g. reporting on something that is outside the control of the service provider is normally of dubious value.

Risks

Customers receiving reports on issues of no significance to them will assume the service provider does not understand their concerns, something that is a common cause of complaint by customers.

if a metric is issued, service provider staff will give priority to improving this aspect of the service, and in some cases will neglect other aspects of the service in order to achieve this. Also, if staff believe that a metric is a measure of their performance, they may become defensive and unwilling to cooperate in the collection of accurate data.
Examples of the unintentional impact of metrics include the following.

- **Targets that are too simple**
  A metric reporting call pick-up speeds may make the staff faster when picking up calls, but may also create a tendency to end a call prematurely to meet the call pick-up target. A balance between call pick-up speed and call-handling quality is normally more effective.

- **Team competitiveness**
  Competition between teams to provide the best service can be productive in generating team spirit and service improvements, unless it becomes excessive and the competitive spirit overwhelms the willingness of the teams to cooperate with each other. For example, reporting fix time targets may tempt a team to refer a problem to another group when it is almost at its fix time target, even if the referral is incorrect. Linking fix time targets to correct and incorrect referrals will reduce this risk.

- **Efficiency goals**
  A metric which simply reports on the number of fixes, without any balancing metric on what type of incidents and problems are being fixed, can encourage the practice of breaking activities down into smaller units, so that the workloads managed by individuals or a team appear to increase. This is a change to the logging practices that may be interpreted as a change to the workload that may prompt incorrect service improvement actions.

Timeliness

Design decisions should take into account the timeliness of metric production. Metrics that can be produced only well after the event they describe are of little use and are generally ignored as irrelevant. The reader needs to see important information quickly. It is better to produce basic (but fit for purpose) metrics at the right time than to delay and produce a set of metrics that are too late to be of use.

Routine production of metrics showing a consistent and steady service over time will typically mean the reader stops being interested in the report, so that when a change in the service does occur, the report is not actually read by those that should act on the information. Some metrics and reports are best produced only when they are triggered by an event. For example, exception reports are produced when a control limit has been breached or are triggered by an alert.

Design features

Keeping metrics simple

If a metric requires a lot of explanation and definition, then it is unlikely to be easily understood (or simple to produce). Simple metrics normally have a stronger impact and it is much easier to engage the interest of the reader. It is also inadvisable to make a report compact by designing metrics that incorporate many variables. Metrics should normally be limited to illustrating the effect of a maximum of three variables to keep the metric easy to understand. Reports that present a partial or confusing picture of the service will damage the credibility of the whole service and of the service provider.

It is also important to avoid the use of IT or service management jargon when designing reports for customers. For example, however important the distinction is between incident and problem in ISO/IEC 20000, the service provider should not assume the difference between incident and problem is understood by customers. More importantly, why should customers be required to adopt their service provider's terminology?

Targets

ISO/IEC 20000 requires targets to be agreed for both the customer's service (in SLAs) and in internal supporting agreements (frequently referred to as operational level agreements (OLAs)). Performance can then be measured against those targets. If there is no agreed target, a service provider may find it useful to use an average service level or workload based on an agreed period of time.

Control limits

ISO/IEC 20000 does not specify the use of control limits, but it can be useful to include the upper and lower limits of acceptable performance in a metric. This often takes the form of a limit being established as an alert or alarm to inform the service provider that the service is approaching an unacceptable level.

Colour coding

Colour coding of metrics, e.g. red-amber-green, is a method of using control limits with a high visual impact. Red coding indicates that something is seriously wrong, amber an alert that there may be a problem, and green that the service is acceptable. Colour blindness or printing/copying in monochrome will obviously reduce the value of this.

Data quality

Understanding the data collection process and the quality of the data collected is of utmost importance. Take, for instance, the time and date of an event. Where has the time and date that has been reported come from? If the time and date is automatically collected (using a performance monitoring system), then the approximations and limits in the system need to be understood, e.g. does the system log events to an exact time (hours, minutes, seconds) or does it group them by time bands (such as within half hour slots)? Does the system ignore certain types of hardware when carrying out an audit of the infrastructure used for configuration management? If the data has been derived from a call logging system (as used by a service desk) then the limits of that system need to be assessed, as well as any potential impact from human error on the data recorded. For example:

- if the events recorded in a logging system are classified according to type of event, then the classification must be supported by simple and unambiguous guidelines. Staff must be trained in the classification process;

- if the location of a caller is logged and a new office opens but the system is not amended to allow the new location to be entered, the quality and value of the information on workloads by location is undermined.

The service provider must ensure that the people logging data used in metrics are properly trained and follow the logging instructions properly. If this is not done, logging practices will vary from person to person, and logged data will be difficult to interpret and may not be usable for metrics production.

Example: Inaccurate data

A service provider received many complaints about the lack of understanding of the customer's business, and also that service reports on serious system failures were incorrect and misleading. A customer satisfaction survey also identified the same causes of low satisfaction.

The service provider and customer had agreed guidelines for setting priorities for fixing errors as part of the original SLA negotiations, which had occurred two years before. Because the complaints were so serious they decided they needed to improve the guidelines and considered re-opening discussions on this with their customers.

At the same time a team carrying out service improvements did some analysis of the mix of priorities logged by members of the service desk. Some very wide variations were found in the proportion of each of Priority 1, 2 and 3. Some of the variety was due to individuals with specialist roles. However, variation occurred within a team of three who all had an identical role, all worked the same hours and who all took calls randomly, as soon as the call came through.

It was established that each of the three had a different view on how to allocate priority. The three pie charts illustrate the differences found in the allocation of priorities.

Analyst A was aware of the guidelines on priority setting agreed during SLA negotiations. Analyst A had been involved in the development of the guidelines.

Analyst A logged a higher proportion of calls as Priority 1 than had been expected when the guidelines were agreed.

Analyst B had less than six weeks experience, having worked for a long time for another organization. All but a small proportion of calls were left at the default of Priority 3.

Analyst C was not aware of the agreed guidelines, even though Analyst C had been in post for 18 months. Analyst C used all three priorities, but with a much higher proportion of Priority 2 calls, compared to Analysts A and B.

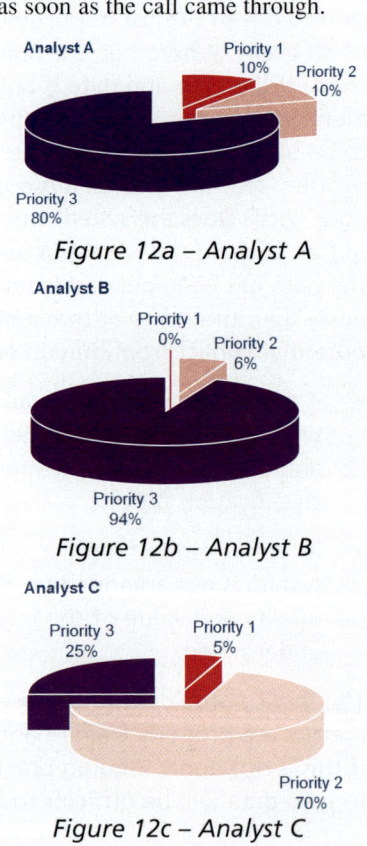

Figure 12a – Analyst A

Figure 12b – Analyst B

Figure 12c – Analyst C

It was not necessary to change the guidelines. Instead they trained all staff in how the existing guidelines were to be used and included this in staff induction. They built the guidelines into the logging system so that they were easily referenced. They also had a programme of awareness on the importance of understanding priorities and the use made of the classification.

The next customer satisfaction survey showed that the customer believed there was an improvement in the service provider's understanding of the customer's business concerns. There were fewer complaints about the quality of reports on serious system failures. The service provider attributed this improvement to the more effective setting of priorities.

Key points:

The service provider realized how important it was to not only understand the customer's business needs, but to have consistent processes and procedures for prioritizing work and classifying events, such as calls to the service desk. They also realized that this should be underpinned by guidelines and training, with regular quality checks on consistency.

Charts versus tables versus text

Although many people prefer charts and diagrams because of their immediate visual impact, some readers who work with numbers as part of their role, such as accountants, prefer tables and lists. Some people prefer text to either charts or tables. Designing a report for a 'typical' user should take into account the preferences of the reader. Most people in the same type of role will have similar preferences for the style of data presentation. The difference in presentation styles is illustrated in the next example, with the same data formatted in a table, pie chart and two types of bar chart.

Example: Illustration of presentation styles

This pie chart shows the call volume per department in a clear and simple style. However, this pie chart is limited in that it does not allow the average to be shown and loses what may be useful extra information from the metric.

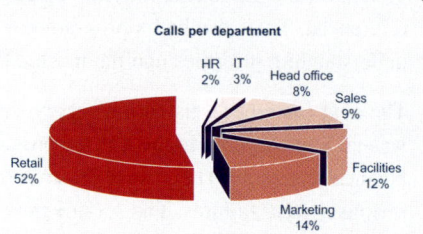

Figure 13a – Metric as pie chart

This version of the data identifies that the largest number of calls is from the retail department. This is useful but of less value than the table or chart below, as it is relatively predictable that retail will make the largest number of calls because it is the largest department. Therefore, this report does not add much insight to understanding the service.

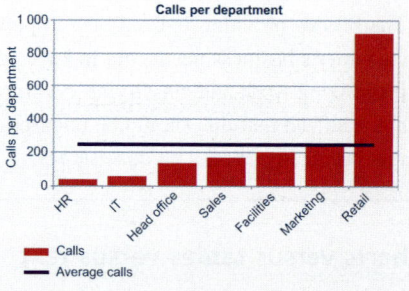

Figure 13b – Metric as bar chart

This version of the data presents the calls per user in each department. This illustrates that Facilities, Sales and in particular, Marketing departments make above average calls per user. This method of presenting the information is useful in identifying important differences between groups of users. This is the first step towards understanding why calls are made and how to prevent calls being made and is, therefore, the first step towards reducing workloads. This chart provides information that is not available in the other charts, but is still simple to understand.

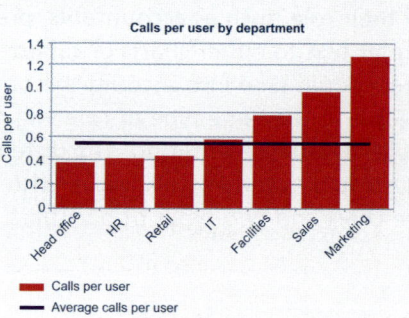

Figure 13c – Metric of calls per user chart

This table shows the basic data in Figures 13a and 13b, but adds value by also showing the average number of calls per user per month for each department. Value has also been added by listing the departments in order of their size. This makes it easier for the reader to identify the largest departments, but makes the other two columns less easy to understand. This form of data representation may be preferred by customers who work with tables of numbers on a day-to-day basis.

Department	Department size	Calls logged	Calls per person
HR	88	36	0.41
IT	96	54	0.56
Sales	165	162	0.98
Marketing	210	252	1.2
Facilities	273	216	0.79
Head Office	379	144	0.38
Retail	2229	936	0.42

Figure 13d – Metric in tabular form

Statistics

Statistical analysis increases confidence in the judgments that are made using service management reports. Those responsible for the design of metrics need to be confident about working with numbers and at least one member of the team must have a good understanding of basic statistical concepts.

Many monitoring and logging tools, as well as spreadsheet functions, will perform statistical analyses. However, using a tool should not be a complete substitute for understanding the strength and weakness of the data, its accuracy and interpretation.

Statistical analysis also helps establish whether what is being reported is a chance variation or a real trend that might require action. Without proper statistical analysis, interpretation will be based on guesswork and assumptions.

It is important that service management decisions are based on verifiable facts and data. For example, if the call volume to a service desk goes up, it may be assumed that there are more customers. Alternatively, the reason for the increase in call volume may be because there is a major problem with the infrastructure that keeps recurring. It needs to be established which scenario has actually happened (possibly both have happened!) in order to know how the situation should be managed.

Although sound statistics should underpin the design of the metrics, the audience for the reports should not normally be expected to have training in statistics in order to understand the reports. Exceptions to this may arise with specialist service provider reports that are used internally, e.g. network management reports.

Testing the design

Once the ideas for the metrics have been developed for each target audience, it is advisable to test the ideas with the help of people from the target audience group that have not previously been involved.

Before starting the checks, make sure that the purpose of each intended metric is clear. Checklist 2 can be used to clarify this.

Checklist 2: Metric design	
What does the reviewer think the metric is showing them? Ask the reviewer to describe what the metric means, in their own words	If there is any uncertainty or ambiguity then try to ascertain the cause, for example is it: • **a matter of data presentation?** How would the reader prefer the information to be presented, i.e. in a chart or table? • **the terms being used?** Has technical language been introduced for a non-technical audience? Has service management terminology been introduced which is not appropriate for the audience? • **about a technical issue that the reader does not understand?** Is this a service provider metric presented as a customer report, or is it a technical subject outside the reviewer's area of expertise and interest?
Is the metric easy to understand or does the reviewer have to spend time thinking about what the report contains?	If the reviewer needs more than a few minutes to absorb the information in any one metric, reconsider the volume of information that is being related in individual metrics. It may be better to provide less information in each metric, separating information in complex metrics into a series of simpler metrics. Alternatively, supplementary text with the metric may save the reviewer time and effort.
Is the information interesting?	Check if the reviewer would like to see the metric on a regular basis, and if so, why. Perhaps, they would like to see the metrics only when events happen or service levels fall outside agreed limits. Note if they think a colleague would find it useful and interesting, and why.

Checklist 2: Metric design (*continued*)	
Is the information useful?	If the reviewer believes that they would find the metric useful, explore how they would use it. This feedback could lead to an improved metric design and a greater understanding of the reviewer's interests and concerns.
Is it obvious from the metric what action should be taken?	This is linked to the previous question. If the information were of use, then how would it be related to the actions the reader would implement once they have seen the metric?
Is the metric better than the existing metrics that the reviewer already receives (if applicable)?	Ask for a comparison with the metrics they see now. If they do not read the existing metrics, ask why not, as this will give insight into what they are interested in and allow the designers to learn from experience.
Does the metric cover the key issues?	This is an open-ended question that needs to be carefully used. Not all metrics of interest can be produced in practice.
If there are any gaps, is this a concern?	If this question is taken as an open invitation to have any metric they find interesting, there may be difficulties later on.

CHAPTER 7

Documenting metrics

The process and procedures followed for the production of metrics and service reports is often not documented, or if it is documented, the documents are not under the control of a change management process. A clear description of each service report is a requirement of ISO/IEC 20000-1. If service reporting documentation is not effective, the production method may vary, and metrics may be misinterpreted when the information in them is used for improvements and process management. The absence of documents describing each service report (and the metrics they are composed of) is a serious risk to the service reporting process. This will normally be checked during an audit.

The documents can cover detail that ranges from the algorithm used to calculate a metric through to a check date when a whole report should be reviewed. It is particularly important that those who use the metrics understand the algorithms that are used in their calculation, to avoid the risk that the metric will be misinterpreted or miscalculated in the future.

Documentation steps

It is important to decide exactly what is going to be measured, and how it will be measured. The following stages might be included in a commonly produced report such as 'percentage of incidents and problems fixed within eight hours'.

- Agree and document the target audience.
- Agree and document the purpose of the metric.
- Define which of policy, process or procedure the metric is reported on.
- Link the metric and the service catalogue entry or entries that the metric relates to.
- If the metric is part of the interface management between two processes (e.g. incident management and problem management, or problem management and change management) then this needs to be recorded.

- Agree the data source (for example a call logging system for incident and problem metrics).
- Agree how the metric is defined.
  - What is included and excluded as an incident or problem in the calculation?
  - Which fields in each record will be used in the calculation?
  - At what times will the service clock start and stop (if not 24x7)?
- Agree either the date when the metric will no longer be required, or when the metric should be reviewed for continued effectiveness.

Once the metric has been defined, anything that affects how it is calculated should be assessed so that the impact of change can be understood and properly managed. For example, if the definition of a Priority 1 incident changes so that more incidents are classified as Priority 1, there will be a change in the metrics that report incident data by priority. In this example, the change in priorities represents a change to the definition of priority, not the characteristics of the incidents, but this may not be obvious unless the relationship between the priority setting guidelines and the actual reported service levels and workloads is understood by the reader.

For many service providers the responsibility for agreeing classifications and the responsibility for the routine production of the metrics is held by different departments, making it even more important that this aspect of changes to metrics is effectively managed.

Documenting changes to metrics

For the most important metrics it may be advisable to track any changes in definition by monitoring and reporting on the difference between the old and new metrics for a set period of time, as well as fully testing the changed metric. This will help identify the implications of the new definition, so that changes arising due to the new classification can be distinguished from other changes.

The metric documentation outlined in Table 3 should be kept under the control of the change management process. Change management is described in more detail in BIP 0035, *Enabling change*.

Table 3 – Metric documentation

Reference
A code should be used so that each metric can be referenced unambiguously and controlled more easily. It may be useful to use a coding system that groups metrics of a similar type, i.e. by policy, process or procedure. The coding system may also be designed to reflect the hierarchical relationship between metrics.
Name
This will simplify discussions during the design and production of metrics with the target audience.
Audience
This is best kept at role/function/department/organization level. The use of the names of individual people raises complications for the currency of the data and should be avoided.
Objective/purpose of metric
The reason for the metric being produced should be briefly defined. This will help during the PDCA review of metrics.
Links/cross-references
Recording a link between the metric, the relevant policy/process/procedure, and the service catalogue entry to which it relates will help define the objective/purpose of the metric. Service providers aiming for ISO/IEC 20000 will also find it useful to identify which clause(s) in ISO/IEC 20000 each metric relates to. Some metrics will relate to more than one clause.
Process interfaces
Metrics may be used to manage the interface between processes (e.g. information from the problem management process used to link problem management and change management, and vice versa). Some metrics are based on data from more than one process, in which case the documentation should link the metric to all the associated processes. An ISO/IEC 20000 auditor will also find this information useful.
Algorithms
The calculation of metrics should be unambiguous. The algorithm should be published in an accessible glossary, or if there is any risk of the metric being misinterpreted the algorithm should actually be published with the metric.

Table 3 – Metric documentation *(continued)*

Data source(s)

Where data is taken from a monitoring or logging system it is advisable to include the field names and a description of the data in each field used and not just the name of the system from which the data is retrieved. This will be used to ensure there is understanding and continuity in the metric. For example, if a metric shows the number of incidents, the data source needs to be defined clearly enough so that only incidents are retrieved, and not problems or changes. Where the data is retrieved using a search, the rules of the search must be included in the metric documentation.

Target/control limits, trend or benchmark value

A target, such as a '95% fix time in eight hours', is common for service reports for SLAs. There may not be a target for the service provider's own internal reports, but the service provider may choose to build in control limits, trends or an established industry benchmark. Whatever is used, this should be documented so that it is used consistently. It should also be noted if a metric is only triggered for issue when the actual value is outside agreed limits.

Limits on accuracy

Data needs to be fit for purpose. Higher accuracy at greater expense may not be cost-effective, but low accuracy may give a false picture. The limits on accuracy need to be estimated and recorded with the metric, and if necessary, published with the metric.

Issue date

The content of reports is often time critical. Producing metrics where the content is out of date by the time the report is issued is a waste of time and money. Issue dates can be conveniently expressed as 'x days after month end' or 'y days after a major incident', rather than actual dates.

Check date

A check date could be used to establish a known date when the content of the metrics is expected to have no value (e.g. a metric produced during a series of time-specific changes). Based on the check date, the service provider could establish if the report should still be produced. For customer reports this could be aligned to the date of SLA reviews. The check date can also be aligned with the PDCA process requirements of ISO/IEC 20000.

CHAPTER 8

PDCA cycle and service reporting

The PDCA cycle, illustrated in Figure 14, applies to service management as a whole as well as individual service management processes. PDCA is essential to achieving ISO/IEC 20000 and best practice service management. The service reporting process is fundamental to the ability of a service provider to carry out the check stage of PDCA, in which each process is reviewed to identify improvements.

Reviewing existing metrics and reports

Does your organization produce very few reports? Perhaps it produces reports that are of no real use in service management, or it produces so many it is difficult to identify which are important. Reports add little value if the purpose of each report is unclear. The risk of this happening is greater where the report production is extensively automated, so that little effort is required in producing them. Under these circumstances too many reports may be produced that cover aspects of the service that are unimportant, and they will distract attention away from what really matters.

The check stage of PDCA requires a review of reports and of the service reporting process itself, as described in Appendix A.

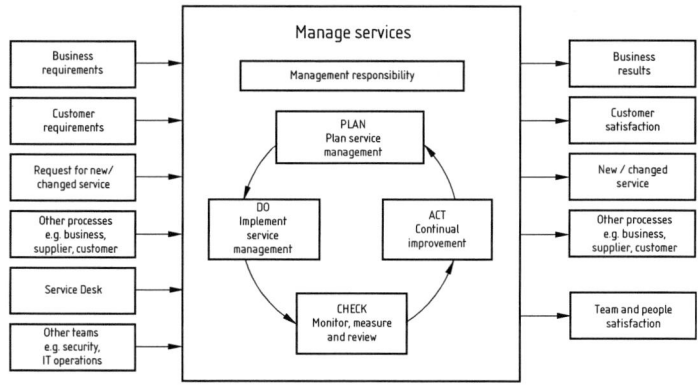

Figure 14 – Plan-Do-Check-Act (PDCA)

Starting point for the PDCA review

As described in Chapter 2 the service reporting process, as for any process in the scope of ISO/IEC 20000, should:

- be directed by policies;

- have a process owner;

- have processes that support the policies being implemented;

- have procedures that define how the service reporting process is carried out in practice.

The PDCA cycle may check the service reports themselves, the documents that describe their production or the service reporting policy, processes and procedures.

The service reporting documents must be sufficient for metrics to be unambiguous and calculated consistently. This is described in Chapter 7, Documenting metrics. Defects in the documents or the service reporting process should be corrected by the PDCA cycle.

Reviewing metrics as part of the PDCA cycle is normally more effective if the metric is linked to one of policies, processes and procedures, as this helps to establish the purpose of each report, whether the report is fit for purpose or whether it must be improved, replaced or stopped.

The PDCA review of metrics and reports should take into account changes in the customer's business activities and priorities, such as the implementation of new technology or services.

It may also become possible to improve existing reports using increased automation or improved reporting tools.

Aspects of PDCA and service reporting that may be audited include:

- how well the purpose of each report is understood;

- whether it achieves its intended purpose (e.g. are improvements identified using metrics and reports?);

- whether it is cost-effective; and

- who uses it.

It is particularly important that the service reporting process provides suitable input to the PDCA cycle, for service improvement planning.

Making improvements to metrics

Producing metrics that work may seem an easy task given that many systems used in service management have reporting functions. However, bad reports may still be produced and this might not necessarily be due to failings in the reporting tool, but other factors such as:

- a poor understanding of what is needed by the target audience;

- bad metric design or metrics so complex that the information they contain cannot be understood or used by the reader;

- insufficient accuracy in the data for the precision required;

- metrics being issued because they are simple to produce and not because they are actually needed;

- metrics being produced that were previously of interest but have no relevance in the current business climate.

Common pitfalls for metrics and service reports that should be identified and corrected by the PDCA cycle include those in Table 4.

Continuity of trend information

When changes to service reports are considered, the continuity of trend information must be taken into account. It is important to avoid misunderstandings arising from issuing what appears to be the same metric, but where the underlying data or algorithms have changed.

Clarity on what changes are made, the reasons for the changes, and the implications for information on trends and forward reporting must all be considered before changes are implemented. For example, changes in classifications such as priority, incident types and change levels may be misinterpreted by the reader as a real change to the service or service levels, whereas the difference is actually due to the new classification system. This should be considered and resolved as part of the change management process controlling the new or amended reports.

Table 4 – Common pitfalls in metric production

Pitfalls	Effect	Solution
Internal diagnostic metrics are presented as performance metrics for the customer.	The customer cannot relate to the information provided by the metrics, so they disregard the reports.	Metrics that are useful only for internal purposes should not be sent to customers. Service providers should meet with customers to find out what they are interested in and how they would use data that meets their needs.
There are so many metrics that irrelevant details mask the important content.	The reports are not considered useful, they cost too much to produce and provide very little value to the customer.	A few key metrics are better, only those that are of real importance to the customer or to the service provider should be produced, and a clear distinction should be made between the two.
The same metrics are reported for many years, without consideration being given to new information needs and changing priorities.	The metrics do not reflect current information needs and so are not read and therefore not used.	The metrics and reports should be reviewed and redesigned to meet the needs of the target audience – in extreme cases, the reports should not be produced at all.
The metrics have no logical relationship to each other, but have been produced by different people at different times.	The information in the reports is confused and therefore not easy to understand.	The reports should be reviewed to give structure and logic to how the metrics are presented – this may well lead to changes in what is reported and will lead to more benefits.
The data is not sufficiently accurate for the metric to be reliable for the management of the service.	The information provided does not act as a firm foundation for actions or management planning. The accuracy of the data is not reliable leading to the wrong conclusions being drawn and the wrong actions taken as a result.	The metrics should be improved in accuracy, or if that is not possible, they should not be issued. If they are issued, they should be accompanied by a warning about the scale of likely errors in the data. The precision with which metrics are reported should also be set so that it does not imply a greater level of accuracy than is justified.
The cost of monitoring and metric production outweighs the benefit of the information provided.	There is a waste of time and money in monitoring and reporting very precise metrics, without an improvement in decision making.	The precision required should be considered so that the accuracy is fit for purpose. A solution like this will require an understanding of how the metrics are used and how the data is obtained.

Table 4 – Common pitfalls in metric production (continued)

Pitfalls	Effect	Solution
Metrics generate bad practices among those who are concerned that 'the numbers' do not reflect well on their personal performance.	Ill-chosen metrics (and ill-chosen targets) damage effectiveness just as much as well-chosen metrics and targets create improvements.	Alternative metrics (and targets) should be selected to encourage the desired practices. Alternatively, the metrics should be supplemented to give a wider picture of performance. For example, a single target for call pick up time can impact the quality of call handling, but a combination of call pick up and call handling quality metrics will give a better view of performance.
Individual teams duplicate effort on measurement and reporting activities.	Wasted effort, conflicts and increased costs of metric production are likely.	The process owner should ensure that the reporting process is streamlined so that duplication is avoided. This may well require centralized control by a senior manager if the individual members of staff are reluctant to give up producing 'their metrics'.
Metrics that are unrelated to the customer's interests are included in the customer's report because nothing else was available at the time it was first produced.	Reports will be irrelevant and customers may see it as an illustration of how little the service provider understands the customer's interests and concerns. At best the report will be ignored, and it may be the subject of complaints.	Discuss reporting requirements with the customer, noting how the customer describes the service themselves, and in what terms they express concerns or requests for new or changed services. Propose new metrics that reflect those interests, then refine the proposed metrics so that they meet the requirements of the customers.
Metrics are manually produced and/or very complex to calculate.	Production is error prone, the results are unreliable leading to inappropriate actions being taken.	Metrics should either be dropped or the production should be automated. If neither option is viable, a simpler method of reporting the data should be sought, but with possible risks from errors in the data identified in the report.
Many metrics and reports are based on small variations, so that a large number of different reports provide similar or the same information.	Production costs are large and the information may be confusing for the service provider.	Discuss the issue with each group receiving metrics and seek a compromise on a standard metric or report.

Example: Reports for monitoring progress improvements

A service provider was making changes to the processes and the service. This was related to the PDCA cycle and to a major change to the customer's business needs. It was expected that there would be a great deal of attention from the media and that failure to deliver on time and to budget would be commented on. The most senior managers in the customer's organization wanted to be kept informed, but they did not want to see a large report or one that included technical details that either they had no interest in or could not easily understand.

The process owner of the service reporting and business relationship management worked with the manager responsible for the continual improvements of service. They designed a dashboard that condensed many, more detailed metrics into three high level measures: cost, timescale and risks. The report was designed to be highly visual, with use of colour and the hemisphere outline shown in Figure 15. This was displayed every week in several locations.

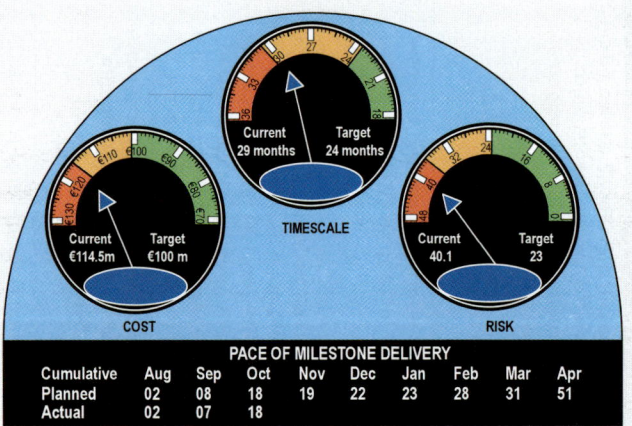

Figure 15 – Dashboard for senior management

Key points:

After some initial problems with the accuracy of the under-pinning data for the dashboard, the report worked well. It resolved the senior management concern that they needed to know what was happening in a style and format they really appreciated. It also improved the communications within the service provider's own organization.

CHAPTER 9

Baselining and benchmarking

When you have established a set of metrics that allows you to produce suitable best practice reports, you are in a position to baseline or benchmark not only your service but also your service management processes. Quantifying the strengths and weaknesses of an existing service is an important precursor to the planning and implementation of service management, the management of major changes and also to continual improvement. Baselining and benchmarking often reveal 'quick win' opportunities that are easy and low cost to implement, providing substantial benefits in process effectiveness and cost reduction.

Baselining

Achieving a thorough understanding of the current service and service management processes is commonly referred to as baselining. Baselining can cover the quality of the service, the service management processes, the workloads that are managed, customer satisfaction and cost-effectiveness of both the service and service management processes.

It is essential that the relationship between the agreed service levels, actual service levels, service costs and the customer's perception of the service quality is understood. Baselining normally includes tracking changes in service quality over time (e.g. has the service got better/worse? Has it become more cost-effective/expensive?). Baselining is therefore repeated at fixed intervals.

It is also advisable to baseline before and after major changes. For example, for new or changed service management processes the cost-benefit can be calculated by comparing the cost and quality of the service before and after the change. Similarly, for major changes to the service, the cost-benefit of the change and the impact on the customer's perception of the service can be measured. Baselining is an important process in ISO/IEC 20000 including the PDCA cycle.

If the service quality has changed, what else has changed? For example, service levels are linked to the workloads, support staff headcount and the extent of automation. A rise in workload, if not planned for in advance, will have an impact on the service levels within weeks or even days. Service levels might be sustained for a short time by the staff reacting by working much harder, but sooner or later the degradation in the service, will be noticed by those reliant on the service.

If an organization is close to meeting the requirements specified in ISO/IEC 20000, a full assessment of the processes against ISO/IEC 20000 may be appropriate. When baselining involves comparison with other organizations, it is usually referred to as benchmarking, as described below.

Benchmarking

Benchmarking compares an organization to an independent reference or standard. This may be a comparison with several organizations, one organization or different units within the same organization. Aspects of an organization that can be compared include the service quality, range of services, value for money, the service management processes or a combination of all of them.

A formal audit against the requirements of ISO/IEC 20000 is a benchmarking of the service management processes and of the management processes (such as PDCA). This is because ISO/IEC 20000 provides an unambiguous best practice model that applies to all service providers, irrespective of size or sector.

Comparisons of actual service levels, rather than service management processes, is outside the scope of ISO/IEC 20000 even though metrics used to compare service levels are produced by the service reporting process. Comparison of service levels may be limited by the need to compare service providers and services that are either the same or very similar for the comparison to be worthwhile.

Many organizations use benchmarking successfully and find that the costs of benchmarking are repaid through the benefits realized from acting on the information provided. Some customers use benchmarking to decide whether they should change their service provider.

It is normal for benchmarking results to differ between service providers. All combinations of service, infrastructure, service provider and customer are unique and most are going through changes. There are also intangible but influential factors, which cannot be quantified directly, e.g. growth rate, goodwill, image and culture. The differences between benchmarking results and the reasons for the differences is often as useful, if not more useful, that the absolute level measured by the benchmarking process. This is because the reasons for differences are normally useful input to planning of service improvements.

Common approaches

Baselines and benchmarks commonly take the following forms.

A point in time

This is an assessment of the same system or department carried out at fixed intervals (usually annual). This is a normal part of service management covered by the PDCA cycle requirements.

Other systems or departments within the same organization

Comparison with other groups in the same organization normally allows a detailed examination of the features being compared, so that it can be established if the comparison is realistic. However, it should be noted that some organizations are unusually diverse and have internal divisions that will differ more than direct comparisons with other organizations.

Industry norms provided by an external organization

Comparison against industry norms provides a common frame of reference, but can be misleading if the comparison is used without an understanding of the differences that exist across a wide variety of organizations.

Direct comparisons with similar organizations

Direct comparisons with other service providers are most effective if there is a sufficiently large group of service providers with similar characteristics. It is important to understand the size and nature of the business area, its sector, geographic distribution, time zones/multinational issues and the extent to which the service is used for business or time critical activities. Benchmarking against a service providers that has different characteristics, such as the comparison of a private sector and public sector service providers, or a retail bank compared to an investment bank, can be misleading.

The culture of the customer also has an influence. Many services are influenced by the extent to which users will or will not accept restrictions on what they themselves may do with the technology provided. For example, it is difficult to have good security processes and procedures with users who will not keep their passwords secure, or who load unlicensed or untested software. In contrast, some users are by the nature of their business activities predisposed to adopt a rigorous approach not only to security but also to accepting the use of standard technology and other methods of reducing the cost of support.

Practical baselining and benchmarking examples

Typical baselining and benchmarking metrics are similar to many metrics used to report against service targets and reports. However they are usually designed with the intention of making comparisons easier by normalizing the data, e.g. costs per unit of storage, instead of simply comparing the total costs of storage and the total volume of storage as two separate metrics.

For quality of process comparisons they can include:

- audits against ISO/IEC 20000;
- results from a self-assessment using BIP 0015, *IT service management – Self-assessment workbook;*
- customer satisfaction reports;
- authorized SLAs as a percentage of the total number of services which should have SLAs;
- proportion of service reviews held;
- rate of errors in a Configuration Management Database (CMDB);
- timeliness of agreed service reporting.

For quality of service comparisons they can include:

- call/request wait time;
- calls/requests per period, i.e. hour, day, week;
- customer satisfaction reports;
- first time fix rate as a percentage of the total fixed (or of the total remote fixes);
- incident/problem resolution time;
- incidents/problems solved per person;
- escalations as a percentage of the number of events;
- remote fix rate as a percentage of the total fixed;
- units of capacity supported per person.

Most benchmarks include some financial metrics (e.g. cost per unit), as assessment of the overall cost-effectiveness is a common reason for benchmarking against other organizations.

CHAPTER **10**

Techniques

Due to the number of techniques used in service management to monitor and report services and service management processes, it is not possible to describe the full range available. The two described in this chapter are useful examples.

Balanced Scorecards (BSC) and Six Sigma®[3] are two popular approaches in the service management industry and are aligned to the approach used in ISO/IEC 20000.

Balanced Scorecards (BSC)

A Balanced Scorecard (BSC) goes beyond the scope of basic IT metrics. It is compatible with ISO/IEC 20000 and can be used as an organizational goal for improvements, covering metrics for all service management processes, interfaces between processes and management system requirements. Although there may be variations in the metrics chosen according to organizational needs and interests, a well-designed BSC gives a rounded view of customers, finances, processes and the skills/abilities of the staff involved. A BSC can enable the service management metrics to contribute towards the full picture for the whole organization.

The BSC approach is inherently top-down. When a BSC is properly designed and implemented, it has a number of features that means it can be aligned with and support a high standard of service management.

The parallels between ISO/IEC 20000 and BSCs exist because a BSC describes what an organization needs to achieve in order to implement its strategy. ISO/IEC 20000 requires all processes to support the implementation of policies.

---

[3] Six Sigma® is a federally registered trade mark of Motorola in the USA.

BSC covers any and all aspects of an organization's business. Although some organizations use more than the four BSC categories shown in Figure 16, most BSCs use four high-level measures that are linked closely to the organization's strategy. Each high-level measure is underpinned by a hierarchy of other measures, providing more details that underpin the level above.

A failure in the BSC approach is often caused by an organization having no real strategy to give them direction. Alternatively, they may have a strategy that has no link to the reality of day-to-day activities and which cannot be used as an organizational goal in practice. BSCs can help managers and staff to develop a set of practical strategies and then link them into what is done on a day-to-day basis, via a hierarchy of metrics. Trying to implement a BSC for the first time is also a test of the way the organization's strategy is described, i.e. are they high-level 'woolly' aspirations or are they suitable as organizational goals.

There are parallels with ISO/IEC 20000. In some cases, the service provider may have many detailed procedures, some processes but few policies. This bottom-up approach leads to accidental overlaps or gaps between processes, failure to deliver against service management policies and a set of documents that do not meet the requirements of ISO/IEC 20000. This is because implementation of processes has been driven by a series of unconnected initiatives, with service managers operating in relative isolation.

If a service provider intends to improve the cost-effectiveness of their service, the development of a supporting BSC requires the service provider to identify what is actually needed, and the method by which improvements should be measured must be agreed before implementation. ISO/IEC 20000 requires a similar logical approach.

Figure 16 – Top-level Balanced Scorecard

Other variations may be used by service providers and can be used as part of the approach to achieving ISO/IEC 20000. For this purpose a BSC would cover the following aspects.

- **Financial:**
  IT budgeting and accounting (clause 6.4);

- **Customer:**
  business relationship management/customer satisfaction (clause 7.2);

- **Ability:**
  management responsibilities (clause 3.1)
  staff competence, awareness and training (clause 3.3);

- **Processes:**
  all processes in ISO/IEC 20000, including PDCA.

The service provider's strategy should be linked to the service management policies and objectives defined in ISO/IEC 20000. Used in this way, the BSC provides a vehicle for leadership and encourages a top-down approach.

The ability scores could be geared simply to the development of skills, but may also cover the willingness to change, as well as the ability to change. In ISO/IEC 20000, the willingness to change is linked to the requirement for management commitment and for staff to understand the role they play in service management, that staff are trained and have the correct attitudes towards customers and delivery of the service. These are all features of an organization that has achieved ISO/IEC 20000 and which has applied the PDCA cycle effectively.

As a methodology, BSC can be combined with Six Sigma, ISO 9000, ISO/IEC 20000, the Deming Cycle and maturity/capability models such as the e-Sourcing Capability Model (eSCM). ISO 9000, Deming Cycle and eSCM are described and compared to ISO/IEC 20000 in BIP 0030, *Management decisions and documentation*.

Six Sigma

The origin of Six Sigma as a measurement standard in product variation can be traced back to the 1920s when Walter Shewhart showed that three sigma from the mean is the point where a process requires correction. Although many measurement standards based on 'a number of sigmas' have been used subsequently, the term 'Six Sigma' in the sense it is used today is generally credited to a Motorola engineer named Bill Smith.

Sigma (which may be represented by the Greek symbol σ) is also referred to as a standard deviation. In Six Sigma the use of a standard deviation is based on the assumption that there will be a range of values above and below a mean, and that the range of results will be a normal distribution (this is often referred to as a 'bell shaped curve'). In Six Sigma, any events that are not within the range defined as acceptable values are defects. The term 'sigma' is therefore used to represent those events that fall at the extremes for normal events, i.e. outside the normally acceptable limits for an event. The number of events that are outside acceptable limits varies depending on what is defined as acceptable, i.e. what 'number of sigmas' has been agreed as the quality objective. This is derived from an algorithm that links acceptable variations, and gives the following dimensions.

- One Sigma = 690,000 defects per million.

- Two Sigma = 308,537 defects per million.

- Three Sigma = 66,807 defects per million.

- Four Sigma = 6,210 defects per million.

- Five Sigma = 233 defects per million.

- Six Sigma = 3.4 defects per million.

There is an inverse relationship between the number of sigma and the number of defects, i.e. One Sigma allows far more defects than Six Sigma but as the examples given show, the relationship is not a simple linear relationship as there are not half the number of defects allowed for Six Sigma as are allowed for Three Sigma.

The philosophy of Six Sigma is to reduce variation in business activities and to support objective, data-driven decisions. In the context of ISO/IEC 20000 'the business activities' are the service management processes, or the individual tasks within the process.

Six Sigma is based on the premise that defects cost money and should be eliminated. Product defects are directly linked to defects in the production methods, and hence a high proportion of defects are considered to be due to defective or inconsistent production methods. When Six Sigma is applied to service management, each variation from an agreed service management process is considered a defect of the process. The logic of Six Sigma is that the variation is a defect even if the variation in the process did not directly impact the quality of the service being managed by the process. This is because the emphasis is on each process being followed more consistently, as well as on improving existing processes by replacing them with better processes.

The total number of iterations, including defects, is needed in order to calculate the proportion of defective iterations of a process. Six Sigma requires a disciplined approach to both the definition of defect and the identification and analysis of the results based on unequivocal data. The definition of the service management processes and the definition of what is a variant/defect become fundamental. This aligns Six Sigma with the requirements of ISO/IEC 20000.

The Six Sigma methodology revolves around the use of a measurement-based strategy that focuses on process improvement and reduced variation through improvement programmes. Measurement and reporting on defects in the service management processes is fundamental to Six Sigma, as it is in ISO/IEC 20000.

The Six Sigma methodology has evolved a vocabulary to cover key points. One of these is DPMO, which is an acronym for 'Defects Per Million Opportunities'. DPMO allows you to take the complexity of a service/process into account by allowing you to choose a target for the acceptable quality level of the process, e.g. the number of defects in the service management process overall or from a single service management process. This has parallels with the use of targets (and 'stretch targets') in the requirements of ISO/IEC 20000 for service management, and in PDCA. Another is DMAIC, which is an acronym for 'Define, Measure, Analyze, Improve and Control'. This robust method can be used to bring improvements on a complex environment.

Six Sigma is increasingly seen as providing benefits to service management. It is used to best effect when creating a new service, when planning to implement a new information system or technology, or when incremental improvement is unsatisfactory. It is a valid approach to implementing and improving service management processes.

CHAPTER 11

Example ISO/IEC 20000 metrics and reports

Example metrics and reports linked to the requirements in ISO/IEC 20000-1 are given in Table 5. This list is not intended to be comprehensive but has been included to illustrate the type of information that is generally beneficial for effective management of services. Many of the metrics and reports are partly text and the list has not been limited to reports such as those produced by the service level management process on actual service levels and workloads.

Many metrics and reports can only be produced by the integration of processes. For example, by information passing between the PDCA cycle and a service management process, and between service management processes (such as incident and problem, capacity and IT budgeting and accounting). For this reason, some metrics could be shown against more than one clause in Table 5, but will only be produced once.

Other example metrics are given in the other publications in the 'Achieving ISO/IEC 20000' series.

Table 5 – Example metrics and reports

Type	Purpose
Clause 3.1 Management responsibilities	
Number/percentage of staff attending communications sessions	Management **shall**:
	b) communicate the importance of meeting the service management objectives and the need for continual improvement;
Staff survey results on quality of management communications	
Customer satisfaction on meeting of customer requirements	c) ensure that customer requirements are determined and are met with the aim of improving customer satisfaction;

Table 5 – Example metrics and reports (*continued*)

Type	Purpose
Actual v planned headcount and costs	e) determine and provide resources to plan, implement, monitor, review and improve service delivery and management e.g. recruit appropriate staff, manage staff turnover;
Percentage of posts vacant, supported by reasons for unfilled posts	
Staff turnover rates	
Review of management processes, especially the effectiveness role of senior responsible owner and processes owners	g) conduct reviews of service management, at planned intervals, to ensure continuing suitability, adequacy and effectiveness
Clause 3.2 Documentation requirements	
Business/customer sign-off requirements specifications, service level agreements and service management plan	Service providers **shall** provide documents and records to ensure effective planning ....
Number or percentage of services and systems adequately documented, including user manuals, operations documents and training materials	Service providers **shall** provide documents and records to ensure effective ..... operation .....of service management
Cost of producing and maintaining user documentation, operational documentation and training materials	
Number of documented policies	...documents and records **shall** include.... a) documented service management policies and plans;
Project/programme management reports	
Percentage of processes documented with correct cross-references between policy, process and procedure	c) documented processes and procedures required by this standard;
Report on changes to a document library, such as new, changed and deleted documents	Quality of document management
3.3 Competence, awareness and training	
Report on the percentage of roles and responsibilities defined, current and correct	All service management roles and responsibilities **shall** be defined and maintained together with the competencies required to execute them effectively
Review results quantified as the percentage that meet requirements	Staff competencies and training needs **shall** be reviewed and managed to enable staff to perform their role effectively
Quality and effectiveness of senior management communications, e.g. from staff surveys etc	Top management **shall** ensure that its employees are aware of the relevance and importance of their activities and how they contribute to the achievement of the service management objectives

Table 5 – Example metrics and reports (*continued*)

Type	Purpose
Frequency of communications on service management from the senior responsible owner or process owner	Top management **shall** ensure that its employees are aware of the relevance and importance of their activities and how they contribute to the achievement of the service management objectives
Clause 4.1 Plan	
Management report on development of a plan (or plans) for service management	Service management **shall** be planned
Clause 4.2 Do	
Financial reports and project/programme reports	The service provider **shall** implement the service management plan to manage and deliver the services, including: a) allocation of funds and budgets;
Project/programme reports	...and including...
Financial reports	f) managing facilities and budget;
Clause 4.3 Check	
Credentials of the proposed auditor	Auditors **shall** not audit their own work
Increase in value and business benefit from implemented changes	Deliver results in accordance with business needs and customer requirements
Increase in frequency and volume of successful changes	Deliver results in accordance with business needs and customer requirements
Number of errors introduced to the live IT services from change, such as major incidents and outages (unavailability) caused by poorly managed change	Deliver results according to objectives in the service management plan
Cost of managing IT assets through better processes and automation, including deployment of IT assets e.g. re-use of software licences	
Decrease in percentage non-compliances e.g. software licences	Deliver results in accordance with policies
Clause 5 – New and changed services	
Sign off by the business and customer that the service and service levels are aligned with the business objectives	The service is deliverable and manageable at the agreed service quality
Customer satisfaction with the service	

Table 5 – Example metrics and reports (*continued*)

Type	Purpose
Percentage variance from estimated costs	The service is deliverable and manageable at the agreed cost
Service delivered within estimated time, resource and costs	
Number of errors introduced to the live IT services from the new or changed service	The service is deliverable and manageable at the agreed cost and service quality
Clause 6.1 Service level management	
Report on the number of percentage of SLAs agreed and in draft status	Each service provided **shall** be defined, agreed and documented in one or more service level agreements (SLAs)
Status of supporting agreements	Agreement of supporting arrangements
Change management reports on changes to SLAs	The SLAs **shall** be under the control of the change management process
Report on review of SLAs with any supporting action plans	Management and control of SLAs
Gap analysis report, with actions	The reasons for non-conformance **shall** be reported and reviewed
Clause 6.2 Service reporting	
All SLA actuals against targets.	Service reporting **shall** include: a) performance against service level targets;
Gap analysis reports	b) non-compliance and issues, e.g. against the SLA, security breech;
Workload volumes reports or exception reports when volumes vary	c) workload characteristics, e.g. volume, resource utilization;
Resource utilization compared to planned or projected utilization	
Customer satisfaction survey results	f) satisfaction analysis
Number and types of complaints	
Action plans included in service reports	Management decisions and corrective actions **shall** take into consideration the findings in the service reports and ......
Management communications on significant events, changes or plans	....**shall** be communicated to relevant parties
Clause 6.3 Service continuity and availability management	
Availability records, linked to a definition of how availability is calculated	Availability **shall** be measured and recorded
Gap analysis or incident/problem management report, plus actions plans	Unplanned non-availability **shall** be investigated and appropriate actions taken

Table 5 – Example metrics and reports (*continued*)

Type	Purpose
Test results	The service continuity plan **shall** be tested in accordance with business needs
Report on recommended action plans	All continuity tests **shall** be recorded and test failures **shall** be formulated into action plans
Clause 6.4 Budgeting and accounting for IT services	
Number/percent budgets produced on time	To budget and account for the cost of service provision
Reports of expenditure against budget	Monitor and report costs against the budget, review the financial forecasts and manage costs accordingly
Standard financial reports, with actuals and targets and any relevant trends or variances	Monitor and report costs against the budget, review the financial forecasts and manage costs accordingly
Financial trends and forecasts e.g. by service, by customer, by business unit	
Budget and costs, with variances for previous month and year to date	Report over and under-spending/recovery
Variation of actual spend against budget for each service	Process that manages the implications of variances against budget
Number of approved changes, by type, that record estimated costs	Changes to services **shall** be costed and approved through the change management process
Number and percentage of planned changes that input into the budget	Budgeting takes into account the planned changes to services
Number of service breaches caused by poor budgeting and cost tracking	Budgeting and cost tracking ensures service can be maintained
Clause 6.5 Capacity management	
Customer sign-off of the business needs input to the development or updating of the plan	Business needs basis for the capacity management plan are based on the customer's views and plans
Baseline measures for each resource type	Status quo baselines as part of the plan
Customer, senior responsible owner and process owner sign-off of the predicted capacity and performance, based on expected changes	Changes to the status quo documented and agreed
Summary of correct and incorrect assumptions	Lessons learned documented and implemented, with management commitment to improvements demonstrated
Underlying cause of incorrect assumptions	

Table 5 – Example metrics and reports (*continued*)

Type	Purpose
Clause 6.6 Information security	
Risk management reports on effectiveness of controls for access	Management of risks associated with access to the service or systems
Change management reports	The impact of changes on controls **shall** be assessed before changes are implemented
Number/percentage of external organizations with defined security requirements	Arrangements that involve external organizations having access to information systems and services **shall** be based on a formal agreement that defines all necessary security requirements
Breaches of security by external organizations	
Number and type of security incidents	Security incidents **shall** be reported.......
Security incident review	Investigation of security incidents
Frequency and types of security incidents	Mechanisms **shall** be in place to enable the types, volumes and impacts of security incidents and malfunctions to be quantified and monitored
Lost service due to security incidents	Cost of security incident(s)
Report with action plans on improvements	Input into a plan for improving the service
Clause 7.2 Business relationship management	
Reports on changes to stakeholder and customer groups	The service provider **shall** identify and document the stakeholders and customers of the services
Report on the review of scope, SLAs etc, possibly as minutes and actions plans	Service reviews
Reports on performance, achievements, workloads, issue logs for the review	
Report on the outcome of the review, possibly as minutes and actions plans	
SLA and contract change details	Management of changes to contract(s), if present, and SLA(s)
Report on complaints and remedial action	Management of complaints
Satisfaction reports, with action plans	Process improvements based on customer satisfaction

Table 5 – Example metrics and reports (*continued*)

Type	Purpose
Clause 7.3 Supplier management	
Types of task or transaction etc., numbers in each type, changes to mix of types	Links to service level management
Details of the reason for a dispute with action plans	Dispute resolution
Cost/benefit for the proposed service improvement	Input to the PDCA cycle and service management processes
Gap analysis and / or risk analysis of know gaps between services provided by each group in the supply chain	Supply chain management
Lead supplier reports on sub-contracted suppliers and service level reports from suppliers	Supply chain management
Clause 8.2 Incident management	
Report on the comparison of number of incident logs and emails, telephone calls and voice mails to front line staff	All incidents **shall** be recorded
Know error reports, problem fixes and CI details	All staff involved in incident management **shall** have access to relevant information such as known errors, problem resolutions and the configuration management database (CMDB)
Major incident report	Major incidents **shall** be classified and managed according to a process
Clause 8.3 Problem management	
Reports on problems by volume, type, technology, cause, impact etc, as regular reports and trends	All identified problems **shall** be recorded
Analysis of the use of classifications and other aspects of logging	Quality check on logging process
Preventive action **shall** be taken to reduce potential problems	Reports on actions for problem prevention
Problem review for individual problem	Problem resolution **shall** be monitored, reviewed and reported on for effectiveness
Problem management process review	
Know error update	Problem management input to incident management
Proposed changes and other improvements	Actions for improvement identified during this process **shall** be recorded and input into a plan for improving the service

Table 5 – Example metrics and reports (*continued*)

Type	Purpose
Clause 9.1 Configuration management	
Number of unauthorized changes	To control the components of the service and infrastructure
Percentage of software and digital media in secure storage referenced from the configuration management database	To control master copies of digital configuration items in secure physical or electronic libraries referenced to the configuration records
Number of non-conformances due to poor control against the root cause e.g. number of unauthorized changes	Configuration audit procedures include the recording of deficiencies, corrective actions and reporting on the outcome
Percentage of components that have data updated and verified automatically	To maintain accurate configuration information
Quality information, including relationships, changes applied, related problems and known errors, status history	Configuration information is visible to those who require it
Clause 9.2 Change management	
Number/percentage of successful changes by root cause	Deliver results in accordance with business needs and customer requirements
Customer satisfaction with each change	
Reduced downtime and errors introduced to the live environment from changes	
Satisfaction with the process(es) (speed, quality, clarity)	Changes have a clearly defined and documented scope
Time to execute a change (through each stage in the change lifecycle)	
Variance from estimated cost of change	
Number of unauthorised changes	The service is deliverable and manageable at the agreed cost
Number and percentage of unplanned and emergency changes	
Clause 10.1 Release management	
Agreed plans for releases based on business plans as well as technical plans	Control of changes and management of risks
Risk management report	Agreement and authorization of plans
Changes to information in the CMDB and details of changes from change records	The release management process **shall** pass suitable information to the incident management process

Table 5 – Example metrics and reports (*continued*)

Type	Purpose
Impact assessment report	Analysis **shall** include assessment of the impact on the business, IT operations and support staff resources, and **shall** provide input to a plan for improving the service
Information on changes input to the CMDB	Integration of release, change and configuration management
Acceptance test results	A controlled acceptance test environment **shall** be established to build and test all releases prior to distribution

Appendix A

ISO/IEC 20000 requirements in summary

It is important to refer to ISO/IEC 20000-1 and ISO/IEC 20000-2 and not rely only on the abstract given here, which only covers those parts of ISO/IEC 20000 that are particularly pertinent to metrics and service reporting.

Other publications in the 'Achieving ISO/IEC 20000' series feature similar tables covering other requirements in the same way.

Each requirement (signified by the use of the verb **'shall'**) is supplemented by informative commentary and recommendations based on the details in ISO/IEC 20000-2 and related publications (see Appendix B).

Table A.1 – ISO/IEC 20000 requirements with informative commentary/guidance

ISO/IEC 20000-1 requirements	ISO/IEC 20000-2 recommendations (*italics*) and additional commentary (**bold**)
Clause 4.3: PDCA Monitoring, measuring and reviewing (Check)	**PDCA is dependent upon the quality of monitoring and service reporting covered by clause 6.2. Service reporting is the single process that crosses every other process, as it is service reporting that is used to manage interfaces between processes. Metrics and reports reflect this cross-process role.**
Objective: To monitor, measure and review that the service management objectives and plan are being achieved.	*The organization should plan and implement the monitoring, measurement and analysis of the service, the service management processes and associated systems.*
The organization **shall** apply suitable methods for monitoring and, where applicable, measurement of the service management processes.	**Reports produced should be used, i.e. production is not an end in itself, and the reports should be of a 'fit for purpose' accuracy.**
	Monitoring needs to be planned so that the right aspects of the service are monitored, not haphazard and based on what is easy to report or what has customarily been done in the past.
	It is important that the monitoring covers areas where there are known concerns, where there is a business critical aspect or where there are going to be changes.
	Suitable automation should be used, especially where this increases the accuracy and speed of the data being collected.

ISO/IEC 20000-1 requirements	ISO/IEC 20000-2 recommendations (*italics*) and additional commentary (**bold**)
	The results of the analysis __should__ provide input to the service improvement programme.
	Items that __should__ be monitored, measured and analysed include:
	a) achievement against defined service targets;
	b) customer satisfaction;
	c) resource utilization;
	d) trends;
	e) major non-conformities;
	f) results of reviews.
These methods __shall__ demonstrate the ability of the processes to achieve planned results. Management __shall__ conduct reviews at planned intervals to determine whether the service management requirements: a) conform with the service management plan and to the requirements of this standard; b) and are effectively implemented and maintained.	**Service reviews, audits, baselining and benchmarking are all linked to PDCA requirements and the production and use of service reports.** **This helps predict what may happen allowing risks and events to be managed effectively. 'Before and after' measurements of services and processes are all important to ISO/IEC 20000, and will contribute to meeting this requirement.**

ISO/IEC 20000-1 requirements

An audit programme **shall** be planned, taking into consideration the status and importance of the processes and areas to be audited, as well as the results of previous audits. The audit criteria, scope, frequency and methods **shall** be defined. The selection of auditors and conduct of audits **shall** ensure objectivity and impartiality of the audit process. Auditors **shall** not audit their own work.

The objective of service management reviews, assessments and audits **shall** be recorded together with the findings of such audits and reviews and any remedial actions identified. Any significant areas of non-compliance or concern **shall** be communicated to relevant parties.

ISO/IEC 20000-2 recommendations (*italics*) and additional commentary (**bold**)

As well as service management activities on measurement and analysis senior management may need to make use of internal audits and other checks.

When deciding the frequency of such internal audits and checks, the degree of risk involved in a process, its frequency of operation and its past history of problems are among the factors that should be taken into account.

In common with external certification and re-certification audits, internal audits and checks should be planned, carried out competently, and recorded as they would be for an independent external audit.

Requirements for documentation as evidence for ISO/IEC 20000 audits are given in clause 3.2 of both Part 1 and Part 2.

Even the most basic of practices will involve recording the objectives of a review and the results of the audit, assessment or review. It is helpful if the method of doing this is linked as closely as possible to the service improvement and remedial actions, i.e. the key findings are directed at what should be done, with the review results as the supporting rationale.

ISO/IEC 20000-1 requirements	ISO/IEC 20000-2 recommendations (*italics*) and additional commentary (**bold**)
Clause 4.4: PDCA Continuous improvement (Act)	**Those aspects of clause 4.4 that are particularly closely linked to service reporting are described below.**

Objective: To improve the effectiveness and efficiency of service delivery and management.

There **shall** be a published policy on service improvement. Any non-compliance with the standard or the service management plans shall be remedied. Roles and responsibilities for service improvement activities **shall** be clearly defined. All suggested service improvements shall be assessed, recorded, prioritized and authorized.	***Planning for service improvements*** *Before implementing a service improvement plan, service quality and levels* **should** *be recorded as a baseline against which the actual improvements can be compared. The actual improvement* **should** *be compared to the predicted improvement to assess the effectiveness of the change.*
A service improvement plan **shall** be used to control the activity. The organization **shall** have a process in place to identify, measure, report and manage improvement activities on an ongoing basis.	*Service improvement targets* **should** *be measurable, linked to business objectives and documented in the plan.* *Service improvement* **should** *be actively managed and progress* **should** *be monitored against formally agreed objectives.*
This **shall** include: a) improvements to an individual process that can be implemented by the process owner with the usual staff resources, e.g. performing individual corrective and preventive actions; and	***Planning and implementing new or changed services*** *Objective: To ensure that new services and changes to services will be deliverable and manageable at the agreed cost and service quality.* *Planning for new or changed services* **should** *include reviewing:* a) *budgets;* b) *staff resources;* c) *existing service levels;*

ISO/IEC 20000-1 requirements

b) improvements across the organization or across more than one process.

The organization **shall** perform activities to:

1) collect and analyse data to baseline and benchmark the organization's capability to manage and deliver service management;

2) identify, plan and implement improvements;

3) consult with all parties involved;

4) set targets for improvements in quality, costs and resource utilization;

5) consider relevant inputs about improvements from all the service management processes;

6) measure, report and communicate the service improvements;

7) revise the service management policies, plans and procedures where necessary; and

8) ensure that all approved actions are delivered and that they achieve their intended objectives.

Major service improvements **shall** be managed as a project or several projects.

ISO/IEC 20000-2 recommendations (*italics*) and additional commentary (**bold**)

d) SLAs and other targets or service commitments;

e) existing service management processes, procedures and documentation;

f) the scope of service management, including the implementation of service management processes previously excluded from the scope.

All service changes **should** *be reflected in Change Management records.*

This includes plans for:

1) staff recruitment/retraining;

2) relocation;

3) user training;

4) communications about the changes;

5) changes to the nature of the technology supported;

6) formal closure of services.

ISO/IEC 20000-1 requirements	ISO/IEC 20000-2 recommendations (*italics*) and additional commentary (**bold**)

Clause 6.2: Service reporting

Objective: To produce agreed, timely, reliable, accurate reports for informed decision making and effective communication.

ISO/IEC 20000-1 requirements	ISO/IEC 20000-2 recommendations (*italics*) and additional commentary (**bold**)
There **shall** be a clear description of each service report including its identity, purpose, audience and details of the data source.	*NOTE The success of all service management processes is dependent on the use of the information provided in service reports.*
Service reports **shall** be produced to meet identified needs and customer requirements.	**The organization should plan and implement the monitoring, measurement and analysis of the service, the service management processes and associated systems. The results of the analysis should provide input to the service improvement programme.** **Service reporting is also important in the PDCA and management of process and organizational interfaces.** *The requirements for service reporting* **should** *be agreed and recorded for customers and internal management. Service monitoring and reporting encompasses all measurable aspects of the service, providing both current and historical analysis.* *Where there are multiple suppliers, lead suppliers and sub-contracted suppliers the reports* **should** *reflect the relationships between suppliers. For example, a lead supplier* **should** *report on the whole of the service they provide, including any services by sub-contracted suppliers that they manage as part of the customer's service.* **Management of the supply chain is described in BIP 0033, *Managing end-to-end service*.**

ISO/IEC 20000-1 requirements	ISO/IEC 20000-2 recommendations (*italics*) and additional commentary (**bold**)
	Purpose and quality checks on service reports *Service reports* **should** *be timely, clear, reliable, and concise.* *They* **should** *be appropriate to the recipient's needs and of sufficient accuracy to be used as a decision support tool.* *The presentation* **should** *aid the understanding of the reports so that they are easy to assimilate, e.g. use of charts.* *Several types of report* **should** *be produced:* *a) reactive reports which show what has happened;* *b) proactive reports, which give advance warning of significant events, thereby enabling preventive action to be taken beforehand (for example reports of impending breaches in SLAs);* *c) forward scheduled reports showing planned activities.*
	The policy on service reporting required by ISO/IEC 20000 should cover when reporting is reviewed and what events might trigger a review (e.g. end of SLA validity period, once a year, before and after a large change). **The role of the process owner is important to ensure that service reporting covers all processes in ISO/IEC 20000 and that it is suitable for the integration of processes and interfaces. This involves cross-checks with other process owners.**

ISO/IEC 20000-1 requirements	ISO/IEC 20000-2 recommendations (*italics*) and additional commentary (**bold**)
Service reporting **shall** include:	**Service providers, suppliers and lead suppliers should produce service reports for their customers and internally for use in management. Part 1 requirements in ISO/IEC 20000 cover generic categories. This list is not prescriptive. A service provider aiming for ISO/IEC 20000 may report different types of metrics, according to their business needs.**
a) performance against service level targets;	
b) non-compliance and issues, e.g. against the SLA, security breach;	**Many of the reports that are useful for process and process interface management are described in more detail in the other publications in the 'Achieving ISO/IEC 20000' series.**
c) workload characteristics, e.g. volume, resource utilization;	*The service provider should produce reports for customers and management covering:*
	a) performance against service level targets, e.g. outage reports, achievements;
	b) non-compliance with standards;
	c) workload characteristics volume information, e.g. incidents, problems, changes and tasks, classification, location, customer, seasonal trends, mix of priorities, numbers of requests for help;
d) performance reporting following major events, e.g. major incidents and changes;	*d) performance reporting following major events, e.g. change, and releases;*
e) trend information;	*e) trend information by period (e.g. day, week, month, period);*
f) satisfaction analysis.	*f) reports that include information from each process, e.g. the number of incidents and the most frequently asked questions, unreliable components of the infrastructure, resource/cost intensive tasks;*
	g) reports to highlight future and scheduled workloads.

ISO/IEC 20000-1 requirements

Management decisions and corrective actions **shall** take into consideration the findings in the service reports and **shall** be communicated to relevant parties.

ISO/IEC 20000-2 recommendations (*italics*) and additional commentary (**bold**)

Process owners play a part in this aspect, especially where actions affect more than one process or if there is a conflict in priorities.

APPENDIX B

Bibliography and further information

Standards

BS 0-3, *A standard for standards — Part 3: Specification for structure, drafting and presentation*

ISO/IEC Directives Part 2, *Rules for the structure and drafting of International Standards*

ISO 9000, *Quality management systems — Fundamentals and vocabulary*

ISO 9001, *Quality management systems — Requirements*

ISO/IEC 17799, *Information technology — Security techniques — Code of practice for information security management*

ISO/IEC 20000-1, *Information technology — Service management — Part 1: Specification*

ISO/IEC 20000-2, *Information technology — Service management — Part 2: Code of practice*

ISO/IEC 27001, *Information technology — Security techniques — Information security management systems — Requirements*

BSI books

BIP 0005, *A manager's guide to service management*

BIP 0015, *IT service management — Self-assessment workbook*

BIP 0008, *Code of practice for legal admissibility and evidential weight of information stored electronically*

PAS 56, *Guide to business continuity management*

Security information

BIP 0070, *Information security compilation on CD-ROM*

BIP 0071, *Guidelines on requirements and preparation for certification based on ISO/IEC 27001*

BIP 0072, *Are you ready for an ISMS audit based on ISO/IEC 27001?*

BIP 0073, *Guide to the implementation and auditing of ISMS controls based on ISO/IEC 27001*

BIP 0074, *Measuring the effectiveness of your ISMS implementations based on ISO/IEC 27001*

Other resources

British Computer Society: www.bcs.org.uk

British Computer Society Configuration Management Specialist Group: www.bcs-cmsg.org.uk

The IT Service Management Forum (itSMF): www.itsmf.com

EXIN: www.exin.nl

Information Systems Examinations Board (ISEB): www.bcs.org.uk/iseb

The Office of Government Commerce: www.ogc.gov.uk

IT Infrastructure Library (ITIL): www.itil.co.uk

BOOKS IN THE
'ACHIEVING ISO/IEC 20000' SERIES

There are ten books in the 'Achieving ISO/IEC 20000' series. Each book in the series includes an abstract of ISO/IEC 20000 that is most relevant to the topic of the book, as well as useful contacts and sources of supporting information. These books can be purchased through the BSI website at www.bsi-global.com.

BIP 0030, *Management decisions and documentation*

This book covers: the background to ISO/IEC 20000; a comparison to other standards and best practice material; compliance and certification audits; the scope of service management; building the business case for achieving ISO/IEC 20000; preparation for an audit and using ISO/IEC 20000 to select your supplier. Important terms that are used in management system standards, where the exact meaning of terms is important to the correct interpretation of the standard, are also explained, including the differences between the terms '**shall**', '**should**' and notes. This book also covers the requirements and recommendations for documents and records, which is a management responsibility requirement in clause 3.2 of ISO/IEC 20000-1.

BIP 0031, *Why people matter*

This book covers the roles and responsibilities of management and process owners, and explains the importance of management commitment to best practice service management, mapping onto the requirements and recommendations of clause 3.1 of ISO/IEC 20000, *Management responsibility*. The book also covers the importance of motivation, training and career development as well as tips and techniques, mapping onto the requirements of clause 3.3 of ISO/IEC 20000-1, *Competence, awareness and training*.

BIP 0032, *Making metrics work*

This book gives a practical view of why metrics and service reports are so important to the delivery of an effective service and to service improvements. It describes the types, the design, target audiences and documentation of metrics used in the service reporting process, covered by the requirements of clauses 4 and 6.2 of ISO/IEC 20000-1, *Plan-Do-Check-Act (PDCA) cycle* and *Service reporting*. Useful tips, techniques and example metrics are included.

BIP 0033, *Managing end-to-end service*

This book describes the supply chains that are commonly managed by service level management, business relationship management and supplier management, which are the requirements in clauses 6.1 and 7 of ISO/IEC 20000-1. It describes the interfaces between suppliers, the service provider and one or many customers. This book also includes useful tips for aspects of end-to-end service, such as the role of service level agreements (SLAs), service reviews, customer satisfaction and complaints procedures.

BIP 0034, *Finance for service managers*

This book covers *Budgeting and accounting for IT services* based on clause 6.4 of ISO/IEC 20000. It introduces financial terms that may be unfamiliar to service management specialists, which will help with understanding the requirements and recommendations. It also covers the relationship between budgeting, accounting and charging, and outlines the importance of service management processes in regulatory compliance.

BIP 0035, *Enabling change*

This book covers the configuration, change management and release management processes which are contained in clauses 9 and 10 of ISO/IEC 20000. It compares the three processes and describes how they interface with each other, and gives advice on the requirements and recommendations of ISO/IEC 20000, example metrics and audit evidence. This book also includes practical advice on meeting the ISO/IEC 20000 requirements on the roles and responsibilities of those involved.

BIP 0036, *Keeping the service going*

This book covers the service continuity and availability management, incident management and problem management processes, which are contained in clauses 6.3 and 8 of ISO/IEC 20000. It explains the role of these processes in keeping the customer's service going, ranging from continuity planning through to the fast-fixing of incidents. It compares the processes and describes how they interface with each other. It includes example metrics and audit evidence, with practical tips and techniques that will help a service provider achieve the requirements.

BIP 0037, *Capacity management*

This book covers the requirements for the capacity management process in clause 6.5 of ISO/IEC 20000. It describes the capacity management process and its role as a link between business plans, workloads, capacity and performance). It also covers the planning required to ensure a service provider is able to deliver a service that allows the customer's business to operate effectively. The book describes capacity management for all types of resources within the scope of service management.

BIP 0038, *Integrated service management*

The opening paragraph of ISO/IEC 20000-1 states that '*This standard promotes the adoption of an integrated process approach to effectively deliver managed services to meet the business and customer requirements*'. This book reflects the importance placed by ISO/IEC 20000 on understanding the interfaces between processes, and how the interfaces are managed so that service management processes are fully integrated. It also reflects the top-down management system approach that is fundamental to ISO/IEC 20000. This book describes how understanding and meeting the requirements of ISO/IEC 20000 gives better control, greater efficiency and opportunities for improvements.

BIP 0039, *The differences between BS 15000 and ISO/IEC 20000*

This book will be of particular interest to those who have used BS 15000 for service improvements, audits or training and need to update their material to reflect the ISO/IEC 20000 standard. ISO/IEC 20000 was based on BS 15000, and this book provides a detailed comparison of ISO/IEC 20000 and BS 15000, for both Parts 1 and 2. It shows the differences in structure, clause numbering and references. The core of this book is a series of tables detailing the changes to the requirements and recommendations clause-by-clause, as well as any re-wording that has been provided to give clarification for an international audience. It includes an explanation of why the changes were made and the implications of each of the changes. This book is based on the material produced by the Project Editor during the drafting of both Parts 1 and 2 of ISO/IEC 20000.